IMPROVE YOUR PHYSICS GRADE

Ronald Aaron
Robin H. Aaron

Northeastern University, Boston

JOHN WILEY & SONS
NEW YORK CHICHESTER BRISBANE TORONTO SINGAPORE

To my wife Marilyn,
     for love and encouragement.

To my late Mother, Mary Aaron,
     who made this possible.
                         ...*Ronnie*

To Mom,
     for love and encouragement.
                         ...*Robin*

Director of Production: Marilyn B. Aaron
Composition: Deborah Peters
Drawings: Peter J. Phillips

ISBN 0 471 89006 5

Printed in the United States of America

10 9 8 7 6 5 4 3 2 1

PREFACE

While other freshmen received letters from home like the following:

Dear Janey,
    How are things? Did you get the homemade prune preserves Grandma sent? Things are fine here. Your little brother bit the mailman again... etc.

                                                Love,
                                              Daddy

I received letters very similar to this:

Dear Robin,

WAVE MOTION

Pay <u>special</u> attention to the difference between wave velocity and particle velocity. This is a <u>tricky</u> subject for most students...etc.

                                              Dad

    My dad is a Physics professor, and his oldest child's first college level Physics course turned out to be a traumatic experience for both of us. My parents became accustomed to frantic calls at midnight, when I would gasp, "Just what <u>is</u> Bernoulli's principle, anyway?" Before exams I would receive at least four calls brimming with last minute test taking tips.

    My father and I decided that it would be a shame if all the brilliant Physics advice of Freshman year should die with the advent of Intro. to Chem I. So Dad and I sat down (for many hours!) to write this book. We are passing our knowledge on to Physics students of the future. We are saving you from some of the frantic fear of your first Physics course. After all, I had an unfair advantage. How many students have Physics professors for fathers? As to those of you who don't - consider yourselves lucky!

SOURCE CREDITS

We are particularly grateful to John Wiley & Sons, Inc. for permission to use figures from *Fundamentals of Physics* (second edition), by D. Halliday and R. Resnick. Copyright (c) 1970. In our book they appear as Figures 6.5, 6.7, 9.1, 10.4, 10.8, 12.11, 14.8(a), 14.9(a), 15.4, 15.10, 17.2, 18.5, 18.6.

This book was also the source for the footnote on page 137, and for Table 8.1.

The "Superman" problem on page 41 is quoted from *University Physics* (sixth edition), by F. W. Sears, M. W. Zemansky, and H. D. Young, Addison-Wesley Publishing Company. Copyright (c) 1949.

Figure 5.1 on page 58 is from *Science Brain Twisters* (Japanese translation), by C. P. Jargocki, Charles Scribner's Sons (Japanese translation arranged through Tuttle-Mori Agency, Inc.). Copyright (c) 1976.

The lyrics to the little "song" on page 136 are by Michael Flanders and Donald Swann. They appear on Angel Record #36388 which is entitled *The Drop of Another Hat*. The particular lyrics that we quote are from the selection "The First and Second Law."

The picture of the old-fashioned meat grinder on page 166 is from a research article by G. R. Plattner which appears in *Few Body Systems and Nuclear Forces II*, edited by H. Zingl, M. Haftel, and H. Zankel. Copyright (c) 1978. Reprinted by permission of Springer-Verlag.

Parts (b) and (c) of Figure 15.2 on page 202 are from *Physics for Scientists and Engineers*, by R. A. Serway. Copyright (c) 1983 by R. A. Serway. Reprinted by permission of Holt, Rinehart and Winston, CBS College Publishing.

Figure 15.6 on page 207 is from *University Physics*, by A. Hudson and R. Nelson, Harcourt Brace Jovanovich, Inc. Copyright (c) 1982.

Figures 18.2 and 18.3 on page 233 are from *Spacetime Physics*, by E. F. Taylor and J. A. Wheeler, W. H. Freeman and Company. Copyright (c) 1963.

CONTENTS

1 HOW TO USE THIS BOOK

Dodo birds once lived on the island of Mauritius in the central Indian Ocean. They were first seen by westerners (Portuguese sailors) in 1507.

The dodo could not adapt to the presence of Europeans. It allowed itself to be caught easily; thousands were slaughtered for food. It carelessly laid a single egg on a palm leaf. These eggs provided tasty food for newly arrived domestic animals such as pigs and dogs. By 1681 the dodo was extinct. The heads and feet of a few dodos are preserved in museums.

Thus ends the sad story of the dodo. It was tame and loveable but not very adaptable, and as a result is no longer with us. Because it reminds us of a large fraction of students in an introductory Physics course, we have chosen the dodo as our symbol. *We shall teach you how to survive in a Physics course, and not to go the way of the dodo!*

1.1   Introduction

The unabashed purpose of this book is to help a typical student improve his or her grade in Physics.   It emphasizes key points and stumbling blocks, and contains test strategies, memorization tricks, and specialized learning techniques.

However, the book is not meant only for students having trouble with Physics problems. We hope that it also will be a valuable supplement for students doing well in their Physics course, who wish to gain further insights into the subject.

An important aspect of this book is the fact that one of its authors is an active researcher who is also interested in education. For this reason, you will glean a feeling and appreciation for Physics, not available in many of the widely used introductory Physics textbooks.   The other author is a sophomore at college, freshly out of an introductory Physics course. She knows the trials and heartbreaks of the beginning Physics student, and understands the need for simple and clear explanations. Also, she is able to point out the snares that she fell into as a freshman dodo.

1.2   Studying Physics—Do's and Dont's

* * * * * * * * * * * * * *

I.   **DON'T** read any "professional" books (Psychology or otherwise) on how to study. They will only depress you! For example, one Psychology text suggests studying by the SQ3R method where SQ3R stands for: Survey, Question, Read, Recite, Review. For effective note taking in class it suggests the LISAN approach. LISAN is an acronym for: Lead, don't follow, Ideas, Signal words, Actively listen, Note taking.

Do you believe that such approaches can help you?

Do you believe in Santa Claus? The Easter Bunny?

2.   **DON'T** accent large fractions of each page of your text with a yellow (or pink) marking pen. You will learn almost no Physics, and at the same time reduce considerably the resale value of your book. After all, who wants to buy a book in which the pages are all yellow (or pink) and completely "studied out"?

If you follow the above suggestions, this book already will have saved you more money that its purchase price. You will have avoided the cost of a "How to Study" book. You will have avoided the cost of an accent pen. And, finally, you will have enhanced the resale value of your Physics text!

DO use the W̲ method for studying Physics. W̲ stands for W̲riting. Let us elaborate:

> The more examples that you do, the sooner things will "fall into place". Each chapter of a Physics text contains several examples. Reproduce them in w̲r̲i̲t̲i̲n̲g̲ on a piece of paper without looking at the text. Keep doing it, again and a̲g̲a̲i̲n̲ if necessary, until there are no errors.

> Most Physics textbooks have separate question sections and problem sections at the end of each chapter. Few professors ever assign these questions, although they help considerably to promote understanding. Answer as many such questions as time permits, in w̲r̲i̲t̲i̲n̲g̲. (You may refer back to the text).

> W̲r̲i̲t̲e̲ down, using complete sentences, any questions you may have concerning the text material, and present them to your instructor. Most professors find it a pleasure to answer well prepared, thoughtful questions.

DO use the D̲D̲ method for effective learning in lecture.

> D̲D̲ stands for D̲on't D̲aydream!

> Studying a chapter (using the W̲ method) before a lecture helps prevent daydreaming.

While, in the above material, we have poked fun at psychologists and professional educators, we and they do have an important goal in common - that is, to get you to participate actively in the learning process. It is generally agreed that writing and reciting aloud constitute active studying, while reading and accenting in yellow (or pink) are passive processes.

> *But be ye doers*
> *of the word, and*
> *not hearers only,*
> *deceiving your own selves -*
>
> James, i, 22

1.3 Using the Book

The material in this book has been designed for use a̲f̲t̲e̲r̲ you have studied your text, attended lectures, and worked hard at the assigned problems. To get the most benefit from the book, we suggest the following strategy:

1. Study chapters in your textbook (using methods of the previous section) before they are covered in lecture. If you cannot bring yourself to study the chapters in detail ahead of time, at least read over them lightly.

2. Don't waste time in lecture sessions writing down material that you know is covered in detail in your textbook. Take notes only when you feel that the lecturer is amplifying or supplementing textbook material. Above all, <u>don't</u> <u>day dream</u> during lectures.

3. Work hard at the assigned problems. Being familiar with the examples in the textbook should get you started in most cases. In the early part of the course, don't be discouraged because you cannot complete many problems, or even start some of them. Most of your fellow students are having the same difficulties.

4. After having followed the preceding study rules (1), (2), and (3), the time has arrived to read and study this book. As in the case of your textbook, be able to reproduce all our examples on paper without error.

Finally, it is important that you attend classes regularly. As long as you concentrate in lecture and recitation sessions you will learn something from them. At the very least, note the material which the instructor emphasizes - <u>it will probably appear on examinations.</u> Also, you learn from other student's questions, and more importantly you must ask your own questions. You are paying the instructor's salary and have a right to ask for clarification. Don't worry about looking <u>Dumb</u>!

1.4 Special Features

Our book is chock-full of memorization tricks and learning techniques. We present some examples below.

I. Problem Categories

The majority of problems encountered in an introductory Physics course fall naturally into groups described by similar equations, similar diagrams, etc. We urge the student to think in terms of such families or groups as a way of reducing the required amount of memorization. It is easier to remember a few categories of problems, than to memorize hundreds of seemingly unrelated problems.

Many modern introductory Physics texts already group their problem sets in terms of related problems. But they seem to do it as a bookeeping procedure only. We emphasize the relations between families of problems as an important aspect of learning the material.

2. Wrong Solutions

There are certain common errors that students make when solving Physics problems. Not making these particular errors would result in considerably higher examination grades. Avoiding common errors is probably the most efficient way of improving your Physics grade!

Along with correct solutions to problems, we discuss the most common mistakes made by students. Some of these errors are so ridiculous, that having been alerted to their existence, you will probably never repeat them. Twenty years of teaching experience shows that without pre-warning, you would probably make such errors.

3.   Calculus and Non-calculus

This book is designed to use with both calculus and non-calculus textbooks. Since many students taking a Physics course have not had calculus, we place all calculus discussions in special sections at the end of each chapter.

Remember, there is no law saying that you cannot use calculus in a non-calculus Physics course! Often the use of calculus will simplify Physics problems tremendously. So, if you feel comfortable with calculus, read our calculus sections - otherwise, ignore them.

2 VECTORS

"Vectors is a subject in mathematics. It is dull and hard to motivate at this early stage in the course. Learn it anyway; otherwise you will get an F for your course grade."

From *The Art of Motivating Students*, by Professor Ronald Aaron (unpublished by popular demand).

2.1 Introduction

Vectors are mathematical quantities that have both magnitude and direction. They are introduced at an early stage in Physics texts, because they can describe important physical quantities such as displacement, velocity, force, etc.

Eventually, vectors make the presentation of Physics clearer, more compact, and more elegant. Working with vectors will help you develop certain mathematical skills and methods for attacking problems that will prove useful throughout your Physics course. So what if the subject of vectors is a bit unmotivated and a bit dull? Nobody promised you a rose garden!

Of course it is important to maintain your sanity. Most students cleverly avoid the danger of being bored and demoralized, by skimming over detailed textbook explanations of vectors. Listening to their instructors and following the diagrams and examples in their books, they eventually get a feeling for the subject, and begin to do problems. We assume that you are at this stage. We shall now try to help you advance your understanding further by giving you a quick and logical review of vectors in two dimensions.

Consider two vectors; we shall call them **i** and **j**.

i is a vector of length one unit which points in the x direction.

j is a vector of length one unit which points in the y direction.

The vectors **i** and **j** are called unit vectors. If you are in a calculus level (typical engineering) course, you will actually use unit vectors in problems. If you are in a non-calculus (typical pre-med) course, you will never use unit vectors (and may skip to Section 2.2). In either case, they provide the neatest way of defining the components of a vector. Any vector **A** may be expressed in terms of the unit vectors by a formula of the form

$$A = A_x \, i + A_y \, j \tag{2.1}$$

A_x and A_y are called the <u>components</u> of the vector **A**.

$A_x\textbf{i}$ and $A_y\textbf{j}$ are called the <u>vector components</u> of **A**.

In Figure 2.1 we display a vector **A** in terms of its vector components. The vector **A** points into the <u>first quadrant</u>.

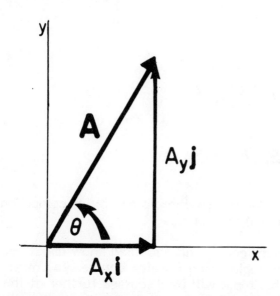

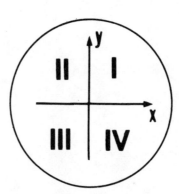

Figure 2.1 The vector **A** points into the first quadrant (quadrant I). The traditional numbering of the quadrants is shown in the circle.

The components of a vector are algebraic quantities. In other words, they are numbers which can be positive or negative, depending on the choice of coordinate system. For example, a vector pointing into the second quadrant would have an x-component which was a negative number; -3 **i** means move three units to the left!

A vector is completely described by its components. However, sometimes an alternative definition in terms of the magnitude and direction angle is more useful. From Figure 2.1 we see that

$$A = \sqrt{A_x^2 + A_y^2} \tag{2.2}$$

$$\tan\theta = \frac{A_y}{A_x}$$

We use the ordinary <u>Roman</u> letter A to represent the magnitude of the vector **A** in the above equation.

The magnitude of the vector **A** is A (or |A|)

In Figure 2.1 the direction angle θ is measured counterclockwise from the positive x-direction (as emphasized by the arrow). We shall always use positive direction angles less than 90º, and generally will indicate by an arrow the direction from which they are measured.

The rules for vector addition are expressed in terms of the components of the vectors. For example, if we add up a bunch of vectors **A, B, C,** etc., to form a resultant vector **R**, we may write

$$R = A + B + C + \ldots \tag{2.3}$$

and

$$R_x = A_x + B_x + C_x + \ldots$$

$$R_y = A_y + B_y + C_y + \ldots \tag{2.4}$$

Equation (2.4) means that you add all the x-components together to get the x-component of the resultant vector **R**, and you add the y-components to get R_y.

If you are taking a typical liberal arts Physics course, all you need know about vectors has been outlined above! If you are taking an engineering Physics course, you must also know something about <u>vector multiplication</u>. This topic will be discussed further at the end of this chapter.

2.2   The Components of a Vector

In order to obtain a high grade in Physics, you must know a small number of simple concepts and mathematical operations absolutely COLD. The answer to certain questions must be almost reflexive, in the sense that if asked the sum of two and two, you would answer "four" without really thinking. The first such idea is the component of a vector expressed in terms of the magnitude of the vector and the direction angle.

For example, if your mother were to awaken you at 2:00 AM, flash Fig. 2.2 in your face, and shout,

"What is the x-component of **A**?"

you must immediately respond,

"A cos θ, Mom!"

If you cannot, you will probably earn a C or even a lower grade in your Physics course.

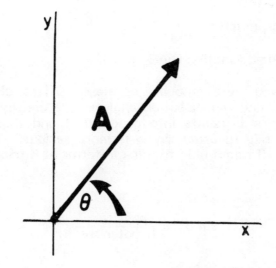

Figure 2.2

Let us now test how well you are prepared to proceed with the study of vectors. At the top of the next page we show a vector **F**. Turn the page and <u>quickly</u> blurt out the x-component of **F**.

GET READY, GET SET, GO!

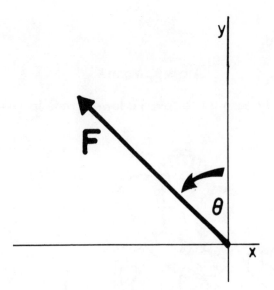

Figure 2.3

Oops! If you answered,

"F cos θ, Mom!",

you need a little more thought, and less pure reflex. First of all, the x-component here is minus F sinθ. Second of all, we're not your mother!

Components are easy, but watch out for these two errors:

1) Wrong sign (minus, plus)

2) Wrong trigonometric function (sin θ, cos θ)

If your algebraic sign was wrong, you were probably careless. After all, the positive x-direction, the positive y-direction, and the direction angle θ are clearly labeled. It is hard to miss the fact that the vector F points into quadrant II and thus has negative x-component. *However getting the trig function wrong is more serious!* You must know simple trigonometry relations COLD. It might help to think in terms of a triangle

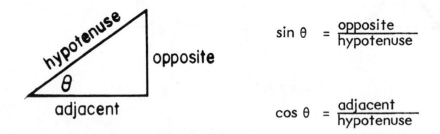

$$\sin \theta = \frac{\text{opposite}}{\text{hypotenuse}}$$

$$\cos \theta = \frac{\text{adjacent}}{\text{hypotenuse}}$$

In other words, learn to associate the word <u>opposite</u> with the sine function, and <u>adjacent</u> with the cosine function. In Figure 2.3 the x-component of F is <u>opposite</u> the indicated direction angle θ, and involves the sine function (not the cosine function).

2.3   Displacement—Vector Addition

When a small object moves from one point in space to another, it is said to undergo a displacement.

Displacement is a vector!

The displacement vector lies along the line joining the two points. Its magnitude is the length of the line, and it is directed from the starting point to the finishing point. As an example, in Figure 2.4 we display three successive displacements (1200 m east, 940 m at 37° south of east, and 1400 m north), along with the resultant displacement vector R. It is possible to obtain R and θ by graphical methods, which involve drawing the three individual displacements to scale in the correct directions, and then measuring the length R and the angle θ.

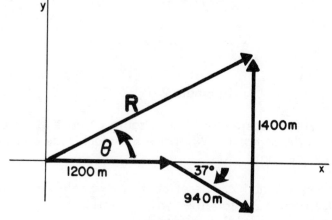

Figure 2.4

Textbooks and instructors (including one of the authors) introduce vector addition with graphical methods, because a hands-on attack with ruler and protractor gives lots of insight to beginning students. However, you must think of the graphical approach as a passing phase in your education (like counting on your fingers and toes) -- grownups don't add vectors graphically! They sum vectors by adding components according to Equation (2.4).

Throughout this book we attempt to develop simple systematic and (hopefully) foolproof methods for attacking problems. We now give you such a rule for vector addition.

> The first step in all vector addition problems is to draw a diagram in which all the vectors act at a common origin. Use this diagram to get the sum.

In other words, instead of using Figure 2.4 to get the resultant displacement, construct Figure 2.5, and use it.

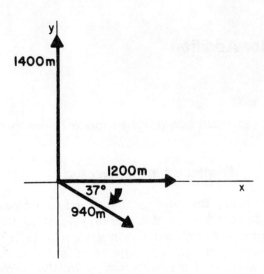

Figure 2.5

* * * * * * * * * * * * * * *

Example 2-1

A hiker walks 1200 m east, then 940 m south of east at an angle of 37°, and finally 1400 m north. What are the magnitude and direction of the total displacement?

Solution: Figures 2.4 and 2.5 describe this example. Forget about Figure 2.4! Use Figure 2.5 together with Equation (2.4) to get the x and y components of the total displacement R.

$$R_x = 1200 \text{ m} + (940 \text{ m}) \cos 37°$$
$$= 1951 \text{ m}$$

$$R_y = 1400 \text{ m} - (940 \text{ m}) \sin 37°$$
$$= 834 \text{ m}$$

The magnitude of the vector **R** is

$$R = \sqrt{R_x^2 + R_y^2}$$

$$= \sqrt{(1951 \text{ m})^2 + (834 \text{ m})^2} = 2122 \text{ m}$$

and the direction angle θ is given by

$$\tan \theta = \frac{R_y}{R_x}$$

$$= \frac{834}{1951} = 0.427$$

$\theta = 23.1°$ (above the positive x-axis)

* * * * * * * * * * * * * *

Example 2-2

Three forces act on an object. A force of 1200 lb acts to the right, a force of 940 lb acts at 37° into the fourth quadrant, and a third force of 1400 lb acts upward. What is the resultant force?

Solution: We hope you have recognized the fact that every step in this example, including the answer, is identical to those in Example 2-1, with the word "meters" replaced by "pounds".

This example emphasizes the fact that our approach to vector addition works for all vector quantities.

* * * * * * * * * * * * * *

2.4   The Sum and Difference of Two Vectors

While we have downplayed the graphical approach for performing vector addition, you should have a certain visual appreciation for the properties of vectors. For example, given two vectors **A** and **B**, you should know that the sum vector **C** = **A** + **B**, and the difference vector **D** = **A** - **B**, lie along the diagonals of a parallelogram. (See Figure 2.6) However, when you are asked to do a calculation, immediately return to the methods of Examples 2.1 and 2.2.

Example 2-3

In Figure 2.6 the vector **A** has magnitude 19, and the vector **B** has magnitude 12. (a) Find the magnitude and direction of the sum vector **C**, and (b) find the magnitude and direction of the difference vector D = **A** - **B**.

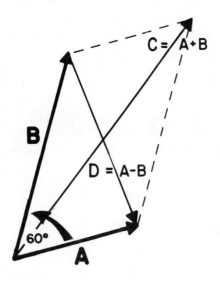

Figure 2.6

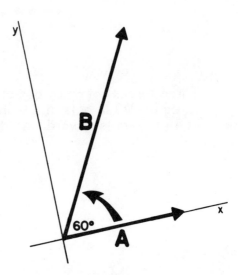

Figure 2.7

Solution: One can't define the direction of a vector without a reference line (coordinate system), so the first step is to choose the positive x-direction. If you wish to do as few calculations as possible, choose this direction to be along the vector **A** (see Figure 2.7). With this choice, only the vector **B** need be resolved into components. We obtain

$$A_x = 19; \qquad B_x = 12 \cos 60° = 6$$

$$A_y = 0; \qquad B_y = 12 \sin 60° = 10.4$$

(a)
$$C_x = A_x + B_x = 25$$
$$C_y = A_y + B_y = 10.4$$
$$C = \sqrt{(25)^2 + (10.4)^2} = 27.1$$
$$\theta_C = \tan^{-1}\left(\frac{10.4}{25}\right)$$
$$= 22.6° \quad \textit{above the positive x-axis}$$

(b)
$$D_x = A_x - B_x = 13$$
$$D_y = A_y - B_y = -10.4$$
$$D = \sqrt{(13)^2 + (10.4)^2} = 16.6$$
$$\theta_D = \tan^{-1}\left(\frac{-10.4}{13}\right)$$
$$= -38.7°$$

The minus sign tells you θ_D is in the second or fourth quadrant. Since D_x is positive, θ_D must be in the fourth quadrant.

2.5 Advanced Vectors

Vectors in Three Dimensions

Geometry with Vectors

Vector Multiplication

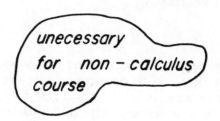

unecessary for non-calculus course

It takes three coordinate axes to describe events in our three dimensional world. Usually these axes are taken to be mutually perpendicular and labeled x, y, and z. The most common convention is to choose the z axis in the following manner: Imagine turning the x axis into the y axis through the smaller angle of intersection (90º); the positive z direction is defined as that direction in which your right thumb would point if you curled the fingers of your right hand in this direction. The resulting coordinate system, illustrated in Figure 2.8, is called a <u>right handed</u> coordinate system for obvious reasons.

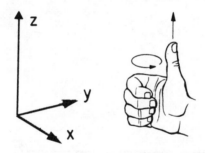

Figure 2.8

With the introduction of a third dimension we need a third unit vector, **k,** which points in the positive z direction. Any vector **A** can now be written in the form:

$$A = A_x i + A_y j + A_z k \qquad (2.5)$$

The vector **A** has three components; A_x, A_y, and A_z. In terms of these, the magnitude of **A** is given by the formula:

$$A = \sqrt{A_x^2 + A_y^2 + A_z^2} \qquad (2.6)$$

Geometry with Vectors

Very few people can easily visualize geometrical relations in three dimensions. Yet, physical events occur in three dimensions, and must be described mathematically. Vectors aid us in such description. We shall not dwell at length on this subject. Rather, we shall give two examples which dramatically illustrate this point, and also will teach you almost everything you need to know about vectors.

* * * * * * * * * * * * * *

Example 2-4

Consider a rectangular box with dimensions 3, 8 and 2 (in arbitrary units).   This box is shown in Figure 2.9 relative to an x-y-z coordinate system. The point P lies at the center of the face ABCD. What is the distance between point P and point E?

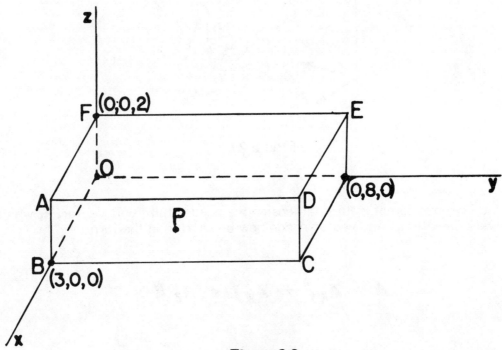

Figure 2.9

Solution:

1.   The vector from the origin O to point P is:

$$r_1 = 3\,i + 4\,j + k$$

2. The vector from the origin O to point E is:

$$r_2 = 0\,i + 8\,j + 2\,k$$

3. The distance from point P to point E is the length of the vector joining P to E; this vector is

$$
\begin{aligned}
R &= r_2 - r_1 \\
&= 8\,j + 2\,k - 3\,i - 4\,j - k \\
&= -3\,i + 4\,j + k
\end{aligned}
$$

The length of this vector is

$$
\begin{aligned}
R &= \sqrt{R_x^2 + R_y^2 + R_z^2} \\
&= \sqrt{3^2 + 4^2 + 1^2} \\
&= \sqrt{26}
\end{aligned}
$$

Note that we have obtained our answer in a straightforward manner, without having to worry about geometrical relationships.

* * * * * * * * * * * * * * *

Vector Multiplication

There are two types of vector multiplication: The cross product which we shall not discuss until Chapter 14 on magnetism, and the scalar product which we shall consider here. The scalar product, as its name suggests, has the same numerical value in all coordinate systems and is thus a scalar quantity.

We define the scalar product, or dot product, of two vectors A and B as:

$$A \cdot B = AB\cos\theta \tag{2.7}$$

where θ is the angle between the two vectors. One can prove that:

$$AB\cos\theta = A_x B_x + A_y B_y + A_z B_z \tag{2.8}$$

18

The fact that the combination AB cos θ can be determined in any coordinate system in which one knows the components of **A** and **B** is very useful. Equation (2.8) allows one to do seemingly difficult problems such as Example 2.5 with ease.

* * * * * * * * * * * * * *

Example 2-5

Calculate the angle between vectors r_1 and r_2 defined in the previous example (example 2-4). This problem looks difficult, but it is really not if you understand the scalar product. Using Equation (2.8) we obtain

$$r_1 \, r_2 \cos \theta = r_{1x} r_{2x} + r_{1y} r_{2y} + r_{1z} r_{2z}$$

$$\sqrt{26}\sqrt{68} \cos \theta = (3)(0) + (4)(8) + (1)(2)$$

$$\cos \theta = \frac{34}{\sqrt{26}\sqrt{68}} = 0.809$$

$$\theta = 36.0°$$

Easy, wasn't it? Solving the problem using ordinary geometry will help you to appreciate the utility of vectors.

* * * * * * * * * * * * *

Note:

You should not think that Example 2.5 is an artificial textbook exercicse. It is the type of problem often encountered by Physicists and Chemists in studying molecular structure.

Note:

The main reason that the scalar or dot product has been introduced is that important physical quantities can be expressed in terms of it. The most famous of these quantities is work, which is the scalar product of force and displacement.

3 MOTION

3.1 Introduction

In this chapter we discuss kinematics, the description of motion in terms of the physical quantities <u>displacement</u>, <u>velocity</u>, and <u>acceleration</u>. For the majority of students this is the first <u>Physics</u> topic in their introductory Physics course. While the subject of vectors is normally covered before kinematics, vectors is a topic in <u>Mathematics</u> (not Physics).

To deal with motion, you must understand the concepts of:

displacement

average velocity

instantaneous velocity

average acceleration

instantaneous acceleration

We shall skip detailed explanations of these topics, because most textbooks discuss them adequately. Rather we shall concentrate on the subject of

**motion in one and two dimensions
with constant acceleration.**

We have researched many examinations, in both "pre-med" and "engineering" Physics courses, in a variety of Universities. We found that, with few exceptions, kinematics problems appearing on exams involved <u>constant</u> acceleration, and much less often, average velocity. We start by discussing average velocity and constant acceleration in one dimension, and then "graduate" to two dimensions.

Variable acceleration (e.g., circular motion or simple harmonic motion) is an important subject, but is more naturally covered along with forces. See Chapters 4 and 9 in this book.

3.2   Average Velocity

The "person off the street" has some notion of the concept of average velocity. If a 200 mile, straight line, trip takes 5 hours, a typical person (with or without a Physics course) would say that the average velocity was the total distance covered divided by the elapsed time, or 40 mph. This is, of course, the correct answer. The Physics definition of average velocity v_{av} is

$$v_{av} = \frac{\Delta x}{\Delta t} \equiv \frac{total \quad displacement}{total \quad time} \tag{3.1}$$

At this point we give you two important tips.

1.   The total displacement is different from the total distance covered. We know of two professors who, in a sadistic fit, gave a problem involving circular motion, and asked for the average velocity for one revolution. The answer is zero (!!) because in one revolution you come back to your starting point. Your total displacement, and thus the average velocity, is zero. No computations are necessary.

2.   Always use the definition in Eq. (3.1). Never try to compute an average velocity by "averaging" velocities. If you stick to the above definition, you won't go wrong on a typical textbook problem on average velocity.

* * * * * * * * * * * * * * *

Example 3-1

This is a typical textbook problem on average velocity: Suppose that you walk 240 feet at a speed of 4.0 ft/s, and then run 240 feet at a speed of 10 ft/s along a straight track. What is your average speed? (The average speed is the magnitude of the average velocity.)

Solution: If you wrote

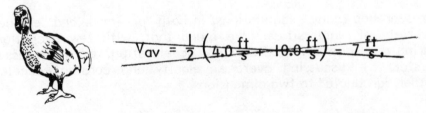

$$v_{av} = \frac{1}{2}\left(4.0\,\frac{ft}{s} + 10.0\,\frac{ft}{s}\right) = 7\,\frac{ft}{s},$$

don't be upset; there is still hope. Just be sure to follow rules in the future.

The correct approach is to note that the time required to walk the first 240 ft is

$$t_1 = \frac{240 \text{ ft}}{4 \text{ ft/s}} = 60 \text{ s}$$

and the time to run the second 240 ft is

$$t_2 = \frac{240 \text{ ft}}{10 \text{ ft/s}} = 24 \text{ s}$$

Using the definition of average velocity in Equation (3.1), we obtain

$$v_{av} = \frac{\text{total displacement}}{\text{total time}} = \frac{480 \text{ ft}}{84 \text{ s}} = 5.7 \frac{\text{ft}}{\text{s}}$$

* * * * * * * * * * * * * * *

3.3 Constant Acceleration in One Dimension

As far as kinematics problems are concerned, it matters little whether you are in a calculus based or non-calculus Physics course. The problems that appear on examinations will almost always involve constant acceleration. Such motion (in the x-direction) is described by the equations,

$$x = x_o + v_o t + \tfrac{1}{2}at^2 \tag{3.2}$$

$$v = v_o + at \tag{3.3}$$

$$v^2 = v_o^2 + 2a(x - x_o) \tag{3.4}$$

where x_o is the position at $t = 0$ and v_o is the velocity at $t = 0$.

MEMORIZE THESE EQUATIONS! Equation (3.4) may be derived by solving (3.3) for t, and substituting $t = (v - v_o)/a$ into (3.2). But Equation (3.4) is used often, and its derivation will cost you lots of time in an examination, SO MEMORIZE ALL THREE OF THE ABOVE EQUATIONS.

Constant acceleration problems come in two common varieties which we refer to as "throw-up" problems, and "catch-up" problems. These are illustrated in Figure 3.1.

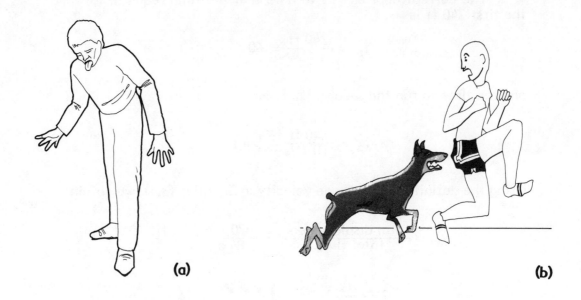

(a)   **(b)**

Figure 3.1   Kinematics problems come in two varieties.
(a) Throw-up problems. (b) Catch-up problems.

We will first study the throw-up variety, because they appear on examinations most often. *The initial step in such problems is to rewrite Equations (3.2) – (3.4), tailored for the case at hand. This means choosing a coordinate system and a starting time, and including the given information.*

* * * * * * * * * * * * * *

Example 3-2

A student stands on the edge of a building 150 ft above the ground and projects an object upward with a velocity of 96 ft/s. It falls to the ground, just missing the building. (Call the vertical axis y.)

(a)   What is the maximum height reached by the object?

(b)   In how many seconds does it reach maximum height?

(c)   What is the acceleration at the maximum height?

(d)   In how many seconds does it reach the ground?

(e)   What is the velocity of the object when it strikes the ground?

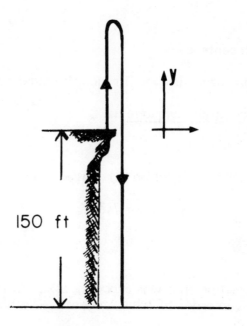

150 ft

Figure 3.2

Solution: We choose the edge of the rooftop as $y = 0$, and the time that the object leaves the ground as $t = 0$. Furthermore we take the positive y direction to be upward.

Thus $y_0 = 0$ and $v_0 = +96$ ft/s. The acceleration is gravity and directed in the -y direction; so $a = -32$ ft/s². Equations (3.2) - (3.4) become:

(I) $y = (96 \text{ ft/s})\, t - (16 \text{ ft/s}^2)\, t^2$

(II) $v = 96 \text{ ft/s} - (32 \text{ ft/s}^2)\, t$

(III) $v^2 = (96 \text{ ft/s})^2 - (64 \text{ ft/s}^2)\, y$

We now give the solution for parts (a) - (e).

(a)   You must know that at the maximum height, the object has zero velocity. Therefore the English sentence

"What is the maximum height?"

translates into <u>equation language</u> as

What is y when $v = 0$?

This means set $v = 0$ in Equation (III) above and solve for y.

$$0 = (96 \text{ ft/s})^2 - (64 \text{ ft/s}^2)\, y$$

$$y = \frac{(96 \text{ ft/s})^2}{64 \text{ ft/s}^2} = 144 \text{ ft}$$

(b) The English sentence

"In how many seconds does it reach the maximum height?"

translates into <u>equation</u> <u>language</u> as

What is t when v = 0?

This means set v = 0 in Equation (II) above, and solve for t.

$$0 = 96 \text{ ft/s} - (32 \text{ ft/s}^2) t$$

$$t = 3 \text{ s}$$

(c) This is a question that was probably asked in the days of Newton and Galileo, in order to trick students. We hope that you didn't answer zero!

The correct answer is -32 ft/s². You know that the object is momentarily at rest at the maximum height. It is the acceleration of -32 ft/s² that starts it on its downward trip.

(d) With respect to our origin, the ground is located at -150 ft, so we wish to find t when y = -150. This means use Equation (I).

$$-150 \text{ ft} = -(16 \text{ ft/s}^2) t^2 + (96 \text{ ft/s}) t \quad \text{or} \quad 16 t^2 - 96 t - 150 = 0$$

Solving the above quadratic equation we get t = 7.29 s. There is also a negative solution which we reject since we are only interested in times greater than zero. The negative solution tells us when the object <u>would have</u> left the ground if it reached the rooftop at t = 0.

(e) To get the velocity with which the object strikes the ground, we substitute the above value of t into Equation (II).

$$v = 96 \text{ ft/s} - (32 \text{ ft/s}^2) (7.29 \text{ s}) = -137 \text{ ft/s}$$

* * * * * * * * * * * * * * *

The Hardest Throw-Up Problem

Consider the following problem from <u>Fundamentals of Physics</u>, by Halliday and

Resnick, Second Edition (Problem 55 on page 41):

"A dog sees a flowerpot sail up and then back past a window 5.0 ft (= 1.5 m) high. If the total time the pot is in sight is 1.0 s, find the height above the top of the window that the pot rises."

Most instructors and students would agree that the preceding problem is the most difficult of the throw-up problems. There are two reasons for this. (1) It seems as if there is not enough information to solve the problem. (2) The essence of the problem is camouflaged by words.

Let us attack the "dog" problem. The first step is to realize that the time it takes the flower pot to pass the window on the way up, is exactly the same time that it takes to pass the window on the way down. So, on the way up, the pot is in sight for 0.5 sec. The next step is to remove the camouflage of words, and rewrite the problem as in Example 3-3.

* * * * * * * * * * * * * *

Example 3-3

The two dots in Figure 3.3 represent photographs of a flowerpot rising in the earth's gravitational field. They were taken at an interval of 0.5 s. How much higher will the pot rise?

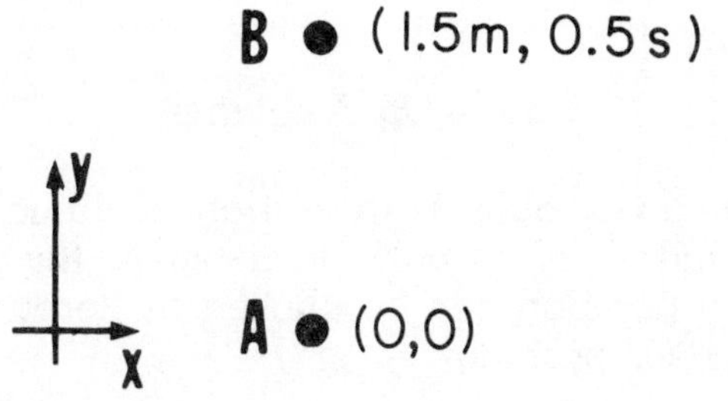

B ● (1.5m, 0.5s)

↑y

+
→ x   **A** ● (0,0)

Figure 3.3

Solution: The crucial step is to choose one of the dots as the origin of a coordinate system (in both space and time). We assign the space-time coordinates ($y = 0$, $t = 0$) to Point A, and choose vertically up as the $+y$ direction. With respect to this coordinate system, the acceleration is -9.8 m/s². The corresponding coordinates of Point B are then ($y = 1.5$ m, $t = 0.5$ s). The constant acceleration equations now take the form

$$\text{(I)} \quad y = v_o t - (4.9 \text{ m/s}^2) \, t^2$$

$$\text{(II)} \quad v = v_o - (9.8 \text{ m/s}^2) \, t$$

$$\text{(III)} \quad v^2 = v_o^2 - (19.6 \text{ m/s}^2) \, y$$

Unfortunately, we cannot use these equations to solve the problem at hand, because we do not know v_o. But we do know that $y = 1.5$ m when $t = 0.5$ s. Substituting these values in Equation (I) gives

$$1.5 \text{ m} = v_o (0.5 \text{ s}) - \tfrac{1}{2} (9.8 \text{ m/s}^2) (0.5 \text{ s})^2$$

$$v_o = 5.45 \text{ m/s}$$

We now obtain the maximum height in the usual way by setting $v = 0$ in Equation (III) and solving

$$0 = (5.45 \text{ m/s})^2 - (19.6 \text{ m/s}^2) \, y$$

$$y = 1.52 \text{ m}$$

Remember that y is measured from the lower dot. Thus the flowerpot rises only 0.02 m above the upper dot.

* * * * * * * * * * * * * *

Catch-Up Problems

Catch-up problems involve two objects in motion, and each of the objects is described by its own set of three kinematics equations. In general, the objects are not at the same position at $t = 0$, so the initial position can be chosen as zero, only for one of the objects. We now present a typical catch-up problem.

* * * * * * * * * * * * * *

EXAMPLE 3-4

A police car is stopped for a red light. A truck moving at a constant velocity of 15 m/s passes the police car, and illegally goes through the red light. Three seconds later, the police car begins accelerating with a constant acceleration of 2.28 m/s² in order to catch-up with the truck.

(a) In how many seconds will the police car catch-up with the truck?

(b) How far will the police car have traveled in this time?

(c) At what time will the police car have the same velocity as that of the truck?

Solution: Each vehicle has its own set of equations. You must do one quick calculation in your head -- when the police car starts moving, the truck is 45 m in front of it -- and then write down the equations of motion for each vehicle, substituting the values of v_o and a.

Vehicle A (police car)	Vehicle B (truck)
$x_{OA} = 0$, $v_{OA} = 0$, $a = 2.28$ m/s^2	$x_{OB} = 45$ m, $v_{OB} = 15$ m/s, $a = 0$
$x_A = (1.14$ m/s$^2) t^2$	$x_B = 45$ m $+ (15$ m/s$) t$
$v_A = (2.28$ m/s$) t$	
$v_A^2 = (4.56$ m/s$^2) x_A$	

(a) The car has caught up with the truck when their x coordinates are identical. So just set $x_A = x_B$.

$$(1.14 \text{ m/s}^2) t^2 = 45 \text{ m} + (15 \text{ m/s}) t \quad \text{or} \quad 1.14 t^2 - 15t - 45 = 0$$

Solving the above quadratic equation we obtain t = 15.7 s (the positive solution).

(b) Substitute the answer from part (a) into the equation for x_A.

$$x_A = (1.14 \text{ m/s}^2) t^2$$

$$= (1.14 \text{ m/s}^2) (15.7 \text{ s})^2 = 281 \text{ m}$$

(c) To find <u>when</u> the velocities are equal, set $v_A = v_B$ and solve for t.

$$v_A = v_B$$

$$(2.28 \text{ m/s}^2) t = 15 \text{ m/s}$$

$$t = 65.8 \text{ s}$$

Oops -- the above answer is <u>wrong</u>! The <u>correct</u> answer is

$$t = 6.58 \text{ s}$$

We purposely made a typical error that students make while under pressure in examinations. When transferring the anwer from the calculator display to the paper, we misplaced a decimal point.

You can defend yourself against such careless errors by <u>thinking</u> about your numerical answers. In this example, the car starts from <u>rest</u>, 45 m behind the truck, and catches up to the truck in 15.7 seconds (if our answer in part (a) is correct). *But our intentionally wrong answer in part (c) implies that the car was going slower than the truck for the first 65.8 seconds.* Thus the answers t = 15.7 s in part (a) and t = 65.8 s in part (c) are inconsistent, and you must conclude that at least one of them is incorrect. In examinations, you will generally receive partial credit for pointing out such inconsistencies, <u>even if you cannot resolve them</u>.

In problems of this type, some students are helped by graphical descriptions as shown in Figure 3.4. The truck is described by the straight line, and the car by the parabola. The vehicles meet when the two curves cross, and their velocities are equal when the slopes of the two curves are equal.

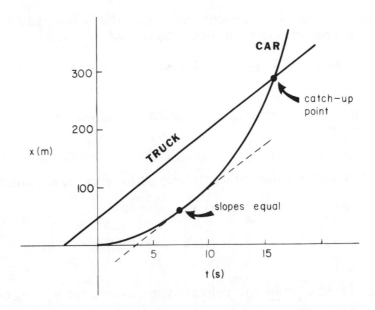

Figure 3.4 The straight line describes the motion of the truck, and the parabola describes the motion of the car.

* * * * * * * * * * * * * *

3.4 Constant Acceleration in Two Dimensions

You might not realize it, but we have already developed the ideas behind two dimensional motion (in our discussions of one dimensional motion). We shall encounter no new equations; yet, the subject is not an easy one. The problem with motion in two dimensions is that you must think in two directions simultaneously. Most students find this difficult, almost like trying to pat one's head while rubbing one's tummy.

Figure 3.5

The most common constant acceleration problems, in two dimensions, involve objects moving freely under the influence of the earth's gravitational acceleration. Such motion is referred to as <u>projectile motion</u>.

IF A KINEMATICS PROBLEM APPEARS ON YOUR FINAL EXAMINATION IT WILL ALMOST CERTAINLY BE A PROJECTILE PROBLEM.

Figure 3.6 represents time lapse photographs of a ball thrown horizontally, with an initial speed of 10 m/s, from the top of a building which is 100 m tall. The pictures were taken at one second intervals. Note that we have projected lines, from the ball to the x and y axes, so that you can follow the path of the ball in both of these directions. This is what we meant when we said that you would have to think in two directions simultaneously. The ball has <u>independent</u> motion in both the x and y directions. We call the actual path of the ball its <u>trajectory</u>.

30

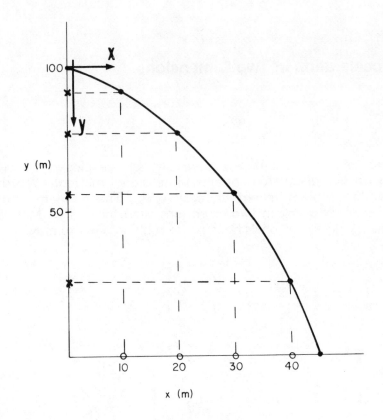

Figure 3.6

We shall develop equations for projectile motion by further examination of Figure 3.6. We deal with the motion in the x- and the y-directions separately, but side by side, in Table 3.1.

x-direction	y-direction
In Figure 3.6, observe the projected lines in the x direction. Every time that a picture was taken, the ball moved the same distance, indicating constant velocity. This is because gravity is the only acceleration present for a body in free flight, and gravity acts in the y direction. Thus the equations describing the x motion are	Gravity acts in the y direction in Figure 3.6. The projected lines in the y direction are like those of a freely falling body. The equations describing the y motion are

$$x = x_0 + v_{ox} t$$

$$v_x = v_{ox}$$

$$a_x = 0$$

$$y = y_0 + v_{oy}t + \tfrac{1}{2}gt^2$$

$$v_y = v_{oy} + gt$$

$$a_y = g$$

Table 3.1

It is instructive to compare Figure 3.6 and Table 3.1. Note that we have taken +y to be the downward direction in Figure 3.6. This choice makes y, v_y and a_y positive quantities, <u>so it is a good choice.</u> If you had made the more traditional choice of +y as the upward direction, the three quantities y, v_y and a_y would all be negative. <u>This is not good.</u> Negative numbers are common sources of error. You must learn a <u>healthy respect for</u> negative quantities. Do not fear them, but be careful when dealing with them, and avoid them whenever possible.

The equations that appear in Table 3.1 apply to all trajectory problems. However, for use in a specific case one must know the constants x_o, y_o, v_{ox}, v_{oy}, and a_y. Let's determine the numerical values of these quantities for the case of Figure 3.6.

By choosing t = 0 when the ball is at the origin, and +y as the downward direction, we determine two of the constants, and the sign of the acceleration.

$$x_o = 0$$
$$y_o = 0$$
$$a_y = +9.8 \text{ m/s}^2$$

You are told that the ball is projected horizontally from the origin, so

$$v_{oy} = 0$$

The determination of v_{ox} is a bit tricky. It must be obtained from Figure 3.6 by staring at the x-axis. Note that the ball is moving "along the x-axis" at a constant velocity v_x = +10 m/s (you are told that the open circles correspond to one second intervals). Since a_x = 0, v_x and v_{ox} are equal:

$$v_{ox} = +10 \text{ m/s}$$

An important element in the solution of all trajectory problems, is determining the constants in the kinematics equations. You will always use some variation of the method that we have just presented.

Important Tip

A typical projectile problem will start off with a statement like,

"At t=0 an object is hurled into the air with an initial velocity **v₀**, which makes an angle θ with respect to the horizontal........"

<u>Your first step</u> in any such problem is to resolve **v₀** into components.

$$v_{ox} = v_o \cos\theta$$

$$v_{oy} = v_o \sin\theta$$

(3.5)

32

Do this for all projectile problems, regardless of the exact nature of the problem. There are two reasons for following this advice:

1. Even Physics professors have the typically human desire to be loved. They don't want to give you a zero on any problem. However by writing gibberish, the typical student forces the issue. Writing down the x and y components of **v** makes it appear as if you know what you are doing, and will almost always earn a few points on a problem.

2. More important, resolving the initial velocity into components, is very often the correct first step in a projectile problem. Wonder of wonders - by following our advice you will be started in the proper direction. Often, the hardest part of a Physics problem is getting started. Once started, one thing leads to another, and who knows?

We now present two examples of projectile problems.

* * * * * * * * * * * * * *

Example 3-5

A soccer player kicks a ball at an angle of 37° from the horizontal with an initial speed of 16 m/s.

(a) What are v_{ox} and v_{oy}?

(b) What are the velocity and acceleration of the ball at the maximum height?

(c) What is the maximum height to which the ball rises?

(d) What is the horizontal distance that it travels?

(e) With what velocity does it strike the ground?

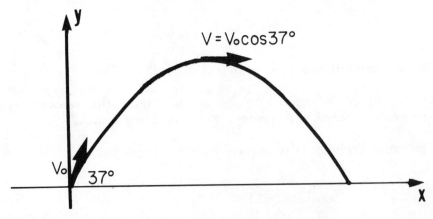

Figure 3.7

Solution: We give the answers to parts (a) - (e) in order.

(a)   If you have read Chapter 2 carefully, you can answer this question immediately and instinctively. From Figure 3.7 we have

$$v_{ox} = (16 \text{ m/s}) \cos 37^\circ = 12.8 \text{ m/s}$$

$$v_{oy} = (16 \text{ m/s}) \sin 37^\circ = 9.6 \text{ m/s}$$

(b)   Were you thinking that "v = 0 at the top"? We hope not!

Remember, velocity is a vector. It has an x and a y component. At the top of the trajectory $v_y = 0$ because the ball has stopped rising, but v_x is equal to its initial value, so

$$v = \sqrt{v_x^2 + v_y^2} = \sqrt{(12.8 \text{ m/s})^2 + (0)^2}$$

$$v = 12.8 \text{ m/s, in the positive x-direction}$$

<u>Needless to say</u>, $a = 9.8 \text{ m/s}^2$ downward!

(c)   "How high does the ball rise?" means find y when $v_y = 0$.

$$v_y^2 = v_{oy}^2 - 2gy$$

$$0 = (9.6 \text{ m/s})^2 - (19.6 \text{ m/s}^2) y$$

$$y = 4.7 \text{ m}$$

(d)   In asking for the horizontal distance covered, we are asking for the value of x when the object strikes the ground -- that is, when y = 0. Let us first find the value of t at which the object strikes the ground.

$$y = v_{oy}t - \tfrac{1}{2} gt^2$$

$$0 = (9.6 \text{ m/s}) t - (4.9 \text{ m/s}^2) t^2$$

$$t (9.6 \text{ s} - 4.9 t) = 0$$

$$t = 0 \quad \text{and} \quad t = 1.96 \text{ s}$$

The ball left the ground at t = 0, and strikes the ground at t = 1.96 s. We now substitute t = 1.96 s into the x equation (remember that there is no acceleration in the x direction).

$$x = v_{ox} t$$

$$x = (12.8 \text{ m/s})(1.96 \text{ s}) = 25.1 \text{ m}$$

(e) The trick here is to remember that the velocity is a vector and has a y-component. The x-component of the velocity is 12.8 m/s because it never changes. The y-component is

$$v_y = v_{oy} - gt$$

$$v_y = 9.6 \text{ m/s} - (9.8 \text{ m/s})(1.96 \text{ s}) = -9.6 \text{ m/s}$$

The magnitude of **v** and the direction angle θ are

$$|v| = \sqrt{v_x^2 + v_y^2} = 16 \text{ m/s}$$

$$\tan \theta = \frac{v_y}{v_x} = -\frac{9.6}{12.8}$$

$$\theta = 37^o \text{ (below the positive x-axis)}$$

* * * * * * * * * * * * * * *

Example 3-6

Suppose that the soccer ball in Example 3-5 is kicked from the edge of a building which is 30 m high.

(a) When does it strike the ground?

(b) How far from the building does it strike the ground?

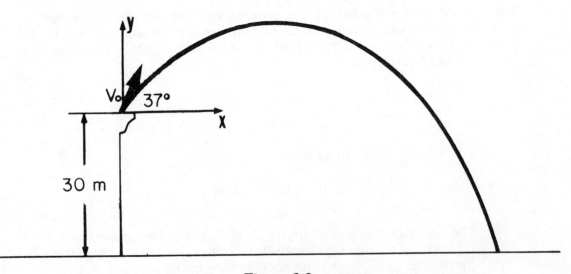

Figure 3.8

Solution: (a) This example is not much different from the previous one. Suppose we take the top edge of the building when the ball is kicked as $(x = 0, \; y = 0, \; t = 0)$. We are asking: What is t when $y = -30$ m (instead of $y = 0$ as in Example 3-5). This means solve the quadratic equation

$$-30 \text{ m} = (9.6 \text{ m/s}) \, t - (4.9 \text{ m/s}^2) \, t^2$$

Answer: 3.64 s

(b) To get the distance from the building, plug the above result into the equation $x = 12.8 \, t$. **Answer: 46.6 m**

* * * * * * * * * * * * * *

Our final projectile example is a fairly common exam question. It involves something falling out of an airplane. The main point is to realize that the object begins its flight with the velocity of the airplane.

In the past such problems have always been warlike (plane drops bomb, etc.), or sexist (man flies airplane, woman never mentioned). We now give an example in which no bomb is dropped, and a woman is the main character.

* * * * * * * * * * * * * *

Example 3-7

A woman is in a jet liner at an altitude of 10,000 ft, travelling horizontally at a velocity of 600 mph (880 ft/s). Attempting to enter the bathroom, she accidentally opens the rear door and steps out into space.

(a) Assuming no air resistance, how long will it take her to reach the ground?

(b) What are the magnitude and direction of the velocity with which she strikes the ground?

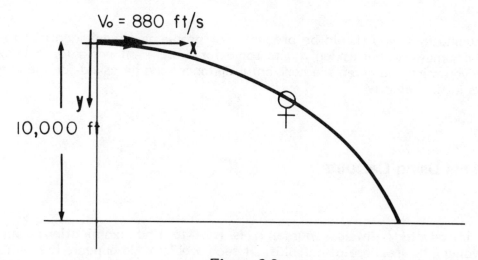

Figure 3.9

Solution: (a) In this case take the woman's starting position as (y = 0, t = 0), and <u>down</u> as positive. Her flight ends when y = 10,000 ft!

$$y = \tfrac{1}{2} g t^2$$

$$10,000 \text{ ft} = (16 \text{ ft/s}^2) t^2$$

$$t = 25 \text{ s}$$

(b) Once again, remember that the velocity is a vector and has an x and a y-component.

$$v_y = g t = (32 \text{ ft/s}^2)(25 \text{ s})$$

$$= 800 \text{ ft/s}$$

$$v_x = v_{ox} = 880 \text{ ft/s}$$

$$v = \sqrt{v_x^2 + v_y^2} = 1189 \text{ ft/s}$$

$$\theta = \tan^{-1}\left(\frac{800}{880}\right) = 42.3^{\circ} \underline{\text{into}} \ \underline{\text{4th}} \ \underline{\text{quadrant}}$$

<u>Don't give the woman an acceleration in the x direction.</u> If you write

$$v_x = 880 - 32 t$$

your instructor will begin to wonder whether his/her life is worth living. However, before making any personal decisions, she/he will give you an F.

$$* \ * \ * \ * \ * \ * \ * \ * \ * \ * \ * \ * \ * \ * \ *$$

Important Tip:

On examinations, you should be prepared for problems almost identical to those you have done for homework, but asking you to solve for a different variable. For example, in Example 3.7 you might be <u>given</u> the horizontal distance, and be <u>asked</u> for the magnitude of the airplane's initial velocity.

3.5 Problems Using Calculus

In an introductory Physics course, it is rare to find examinations which contain problems involving the real use of calculus. If such problems do appear, they will be among the easiest on a particular subject.

* * * * * * * * * * * * * *

Example 3-8

The position of an object is given by the equation

$$x = 5 \text{ m} + (5 \text{ m/s}^2) t^2 - (0.15 \text{ m/s}^4) t^4$$

What are the velocity and acceleration at t = 2 s?

Solution: Differentiate twice.

$$v_x = \frac{dx}{dt} = (10 \text{ m/s}^2) t - (0.60 \text{ m/s}^4) t^3$$

When t = 2 s, v_x = 15.2 m/s.

$$a_x = \frac{dv_x}{dt} = (10 \text{ m/s}^2) - (1.80 \text{ m/s}^4) t^2$$

When t = 2 s, a_x = 2.80 m/s².

* * * * * * * * * * * * * *

You must admit that the above problem involves the real use of calculus, and is very easy. On the other hand, sometimes problems involve an understanding of calculus, but never require that you actually differentiate or integrate. These can be quite tricky. Such an example follows:

* * * * * * * * * * * * * *

Example 3-9

A sky diver of mass m is falling vertically near the earth's surface under the influence of a controlling differential equation

$$m \frac{d^2y}{dt^2} + R \frac{dy}{dt} + m g = 0$$

The sky diver eventually reaches a terminal (constant) velocity. What is this velocity?

Solution: Within a year you will all know how to solve the above differential equation. But you do not have to solve the differential equation to answer the question. You need merely know that

$$v = \frac{dy}{dt}$$

$$a = \frac{dv}{dt} = \frac{d^2y}{dt^2}$$

We then rewrite the differential equation as

$$m\,\frac{dv}{dt} + R\,v + m\,g = 0$$

When v reaches a constant velocity which we call v (terminal), the acceleration a = 0. Substituting these results into the equation above we have

$$m\,(0) + R\,v\,(\text{terminal}) + m\,g = 0$$

$$v\,(\text{terminal}) = -\frac{mg}{R} \qquad \text{(the minus sign means downward)}$$

This seems like a tough problem, but it actually appeared on the final exam of an eastern state university. Presumably the professor discussed similar problems in class, and the students did not find the problem "all that difficult".

* * * * * * * * * * * * * * *

4 FORCES

4.1 Introduction—Newton's Laws

The material in this chapter is designed to accomplish two, main purposes:

1. To establish a systematic approach for attacking all force problems.

2. To emphasize the fact that force problems fall into a few general categories. The trick is to learn the _types_ of problems in each of the few categories, instead of the hundreds of problems encountered in a typical general Physics course.

To obtain these goals we shall develop methods based on _conscious_ application of

Newton's Three Laws:

1. A body will move in a straight line with constant speed unless acted upon by external forces.

2. The vector sum of all the forces acting on a body is directly proportional to the mass of the body, and to the acceleration of the body. In other words:

$$\sum \boldsymbol{F} = m\boldsymbol{a} \qquad (4.1)$$

or

$$\boldsymbol{F}^{net} = m\boldsymbol{a}$$

In everyday conversation we often refer to Newton's Second law as "F = ma". This is poor practice. The Second Law as stated in Equation (4.1) says that it is the vector sum of the forces ($\sum F$) or, equivalently, the net force (F^{net}) that is equal in magnitude and direction to the product of mass and acceleration.

3. Forces are always exerted in _pairs_; for every action there is an equal and opposite reaction. More precisely, if body A exerts a force on body B, then body B exerts an equal but opposite force back onto body A.

Most force problems in an introductory Physics course involve direct application of Newton's Second Law. The First and Third Laws play a less obvious role, but you must understand them! The Second Law defines a force as a physical quantity which causes acceleration. If there is acceleration, there must be a force, and vice versa. Learning to deal with forces, and with Newton's Second Law, is a central feature of success in a Physics course. So, WAKE UP! BE ALERT! At the very least, you must master the material in this chapter if you wish to earn a good grade in a Physics course.

4.2 Some Relations Between Force and Acceleration

Relation I: LARGE changes in velocity in small time intervals can cause LARGE net forces.

Relation I is explained in the following way. Consider the average net force. This is defined in terms of the average acceleration by Newton's Second Law:

$$F_{av}^{net} = m a_{av} \tag{4.2}$$

By definition

$$a_{av} = \frac{\Delta v}{\Delta t} \tag{4.3}$$

therefore,

$$F_{av}^{net} = m \frac{\Delta v}{\Delta t} \tag{4.4}$$

Observe the above equation. A LARGE change in velocity (Δv) in a small time interval (Δt) produces a LARGE force. This principle is used by automobile manufacturers in constructing safer cars. The goal is to decrease the force in a collision by making the collision last as long as possible. This can be done by building the front of the car so that it collapses in a controlled manner under impact, thus extending the time of the collision. Seat belts and padded dash boards perform similar services.

Figure 4.1 The safest car in the world?

* * * * * * * * * * * * * * *

Example 4-1

Problem #30 in Chapter 3 of Sears, Zemansky, and Young, <u>College Physics</u>, reads as follows:

"A student determined to test the law of gravity for herself walks off a skyscraper 300 m high, stopwatch in hand, and starts her free fall (zero initial velocity). Five seconds later, Superman arrives at the scene and dives off the roof to save the student.

What must Superman's initial velocity be in order that he catch the student just before the ground is reached?

What must be the height of the skyscraper so that even Superman can't save her? (Assume that Superman's acceleration is that of any freely falling body.)"

<u>Do not solve the above problem.</u> Rather, explain why the above problem is unrealistic. (In the picture on the following page, we give our impression of what happens when Superman catches the student.)

Solution: Superman's behavior is unrealistic because, besides being very strong, he is also very smart. He knows that he should not catch the student "just" before she hits the ground. In the real world, Superman would reach the student well above the ground, and slow her down over a long time interval. Thus Δt would be large, F^{net} would be small, and the student would reach the ground in comfort.

* * * * * * * * * * * * * * *

Example 4-1

Relation 2: You must associate accelerations and forces with changes in directions of velocities, as well as with changes in magnitudes.

Consider a boy running to the right at 10 mi/hr. Suddenly, he sees a snarling Doberman Pincher ahead of him. Within seconds he reverses his direction, and runs to the left at 10 mi/hr.

Figure 4.2

Therefore (as discussed earlier), there is a large average force exerted on the boy <u>by the ground</u> (see middle segment of cartoon). It is important to realize that Δv is not zero in this situation; the initial and final velocities have opposite sign.

$$\Delta v = v_{final} - v_{initial}$$

$$= 10 \, mph - (-10 \, mph) \quad [\textit{not} \ 10 - 10 = 0!]$$

$$= 20 \, mph$$

Motion on a Curved Path

An object moving in a curved path experiences a force due to the <u>change in direction</u> of its velocity. In an introductory Physics course, the curved paths that <u>you deal with</u> will almost always be circular. When you see the words, "Constant speed in a circle"

DON'T say "Ahh, <u>constant</u> speed -- that means no net acceleration, and therefore no net force."

DO day "Ahh, <u>circle</u> -- that means velocities are changing directions and, therefore, there are accelerations and forces present."

* * * * * * * * * * * * * *

Example 4-2

In which of the two cases described next does the person involved experience serious discomfort.

(a)   In a supersonic jet moving smoothly, straight ahead, through the air at a speed of 1200 miles per hour.

(b)   Moving in a circle of radius 10 meters at a constant speed of 100 meters per second.

Solution: In part (a) the person experiences no discomfort because his/her velocity is constant.   Constant velocity means no change in velocity; no change in velocity means no acceleration; no acceleration means no force. Hence there is no discomfort.

In part (b) the person feels no discomfort because he/she is dead! We are dealing with circular motion; there is an acceleration

$$a \quad = \quad \frac{v^2}{r} \quad = \quad \frac{(100 \text{ m/s})^2}{(10 \text{ m})} \quad = 1000 \text{ m/s}^2$$

This acceleration is about 100 times that of gravity, so the person is experiencing a force 100 times his weight. No one can withstand a force of 100 times one's weight for very long, and remain alive.

* * * * * * * * * * * * * *

4.3   More on the First and Third Laws

As mentioned earlier, most introductory Physics problems require direct application of Newton's second law, while at the same time, involving the first and third laws in less obvious, but <u>important</u> ways.   Following, we give two examples to reinforce your understanding of these laws.

* * * * * * * * * * * * * *

Example 4-3

As a pendulum reaches the lowest point in its path, the string is cut. What is the most probable path that the pendulum bob will follow?

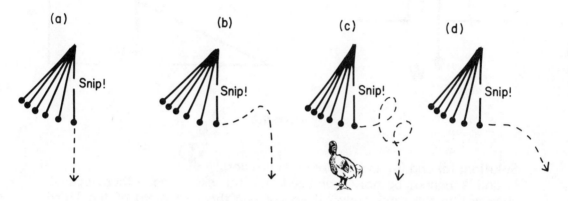

Figure 4.3

Solution: When the string is cut, the pendulum bob has a velocity which is in the horizontal direction. There are no forces in this direction, so Newton's First Law tells us that this velocity component will remain constant. However, the gravitational force is acting downward, so the velocity component in this direction begins to grow.

The correct answer is (d).

* * * * * * * * * * * * * *

Example 4-4

In Figure 4.4 (a) we show a block resting on a horizontal surface. There are two forces acting on it: the force **W** exerted by the earth, and the force **N** exerted by the surface, perpendicular to itself. In Figure 4.4 (b), the same block rests on an inclined plane. In this case there is a frictional force **f** in addition to **W** and **N**.

The reaction force to **W** is

    (a)   the force **N** in both Figs. 4.4 (a) and 4.4 (b).
    (b)   the force **N** in Fig. 4.4 (a) only.

(c)   the force N in Fig. 4.4 (b) only.

(d)   none of the above

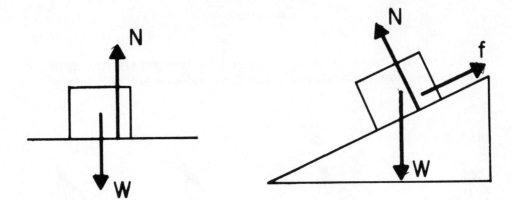

Figure 4.4

Solution: (a) and (b) and (c) are dodo answers!
N and W cannot be action and reaction forces to each other because they act on the same body. If you reread the statement of the Third Law, you will see that the action and reaction forces act on <u>different</u> objects.

So "(d)" is the correct answer. More precisely: The earth exerts a force of magnitude W, in the downward direction, on the block. Thus, according to Newton's Third Law, the block exerts a reaction force on the earth which has magnitude W and is directed upwards. This statement holds for both Figs. 4.4 (a) and 4.4 (b).

4.4   Rules for Solving Force Problems

1.   Draw a Force Diagram   (Free-body Diagram)

2.   Choose a Coordinate System

3.   Apply Newton's Second and Third Laws

In an introductory Physics course, most problems that you encounter can be solved with a systematic approach following a set of rules. The above rules are not very original -- most books and instructors suggest similar approaches. However in the beginning you should follow the rules to the letter.

The most important step in the solution of a force problem is construction of a neat and accurate force diagram, or <u>free</u> <u>body</u> <u>diagram</u>.

Each body of a composite system must be isolated, and all forces acting <u>on it</u> must be drawn and clearly labeled. Be sure that you have included <u>all</u> the forces which act <u>on</u> the bodies in the system in your diagrams. Be just as sure that you have not included any fictitious forces (forces that are not really present).

A good force diagram usually results in a solution which is basically correct. An incorrect force diagram <u>or no force diagram</u> invariably leads to disaster. For this reason we stress the force diagram in the following examples. We also give some dodo diagrams to illustrate common errors made by students. These diagrams are particularly instructive and should be studied carefully.

* * * * * * * * * * * * * *

Example 4-5

An object of mass M slides down a plane inclined at an angle θ. The coefficient of kinetic friction between block and plane is μ. What is the acceleration down the plane?

Correct Approach

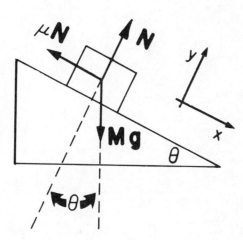

Approach

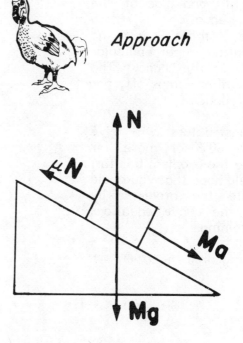

The above diagram is beautiful and contains all the forces. The earth exerts the weight which acts vertically downward. The plane exerts a normal force perpendicular to itself, and a frictional force parallel to itself. And that's it! There are no other forces.

The above is clearly the work of a demented mind. Never mind that the normal force **N** is not normal to the surface; <u>more terrible</u> is the force <u>labeled Ma</u>.

* * * * * * *

Whenever possible one of the coordinate axes should be chosen to be along the line which the acceleration acts, preferably in the direction of the acceleration. Thus we have chosen the positive x direction down the plane. Using this coordinate system, Newton's Second Law yields the equations:

$\sum F_x$:

$$Mg \sin \theta - \mu N = Ma$$

$\sum F_y$:

$$N - Mg \cos \theta = 0$$

The right hand side of the second equation is zero because there is no acceleration perpendicular (normal) to the plane. The block neither jumps off, nor falls through.

Most instructors will give 50% to 60% or more for carrying the problem this far. We would hope that you could complete the problem, by solving the above equations and obtaining

$$a = g(\sin \theta - \mu \cos \theta)$$

A student who commits this atrocity usually explains himself by saying,

"But Ma is the force pushing the block down the plane."

Hopefully, you the reader are aware that it is the x component of Mg that is doing this job!

You must take this example very seriously. The normal force in the wrong direction, and a fictitious force labeled Ma are two of the most common errors in force problems.

* * * * * * * * * * * * * *

Example 4-6

Two blocks, one of mass M and one of mass m, are in contact on a horizontal frictionless table. A horizontal force F is applied to the block of mass M as shown in Figure 4.5.

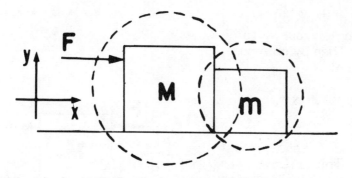

Figure 4.5 Isolate the two objects enclosed by the dashed circles, and draw free body diagrams.

If F = 57 N, M = 65 kg, and m = 30 kg, find

(a) the acceleration of the system.

(b) the force which the object of mass M exerts on the object of mass m.

Correct Approach

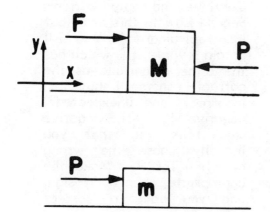

Note that we have omitted all forces acting in the y direction, because they are not really relevant to this problem. We feel that it is OK to omit irrelevant forces if such action makes the diagram clearer. On the other hand, your professor

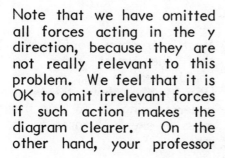

Approach

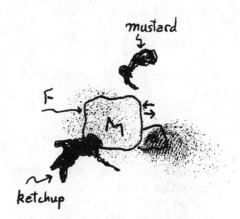

Remember, neatness is next to Godliness. The above diagram is unsatisfactory because it is sloppy <u>and</u> <u>ambiguous</u>.

It is not clear to us, which force is acting on which body.

50

might insist that <u>all</u> forces appear on your force diagrams! Then by all means, include them!

* * * * * * *

You should realize that Newton's Third Law has already been incorporated into the free-body diagrams. Object M exerts a force P on object m. Newton's Third Law tells us that object m exerts a force on M which has magnitude P and acts to the left.

* * * * * * *

We now apply Newton's Second Law to the original composite diagram (Figure 4.5) and to each of the free-body diagrams. The composite diagram yields an equation for $\sum F$ in the x-direction:

(A) $F = (M + m) a$

The free body diagrams give

(B) $F - P = Ma$

(C) $P = ma$

Only two of the above equations are independent, since one can get Eq. (A) by adding Eqs. (B) and (C). This means that we can never solve for more than <u>two</u> variables.

Why have we written down three equations when we only need two? The reason is that, depending on the given information, it is easier to use one equation than another.

* * * * * *

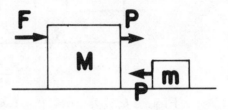

The above diagram is neat and unambiguous. Unfortunately it is <u>also incorrect</u>.

It is incorrect because it implies that pushing mass M to the right with a force **F**, will cause mass m to help the pushing process by providing an additional pull P to the right.

But neatness will pay off. Continue, and apply Newton's Second Law to this neat (but incorrect diagram). You will obtain a value of P which has the same magnitude as the correct answer, but an incorrect (and unexpected?) negative sign. The negative sign tells you that you initially chose the wrong direction for P. Make the appropriate change in sign, and you have a correct solution.

* * * * * *

Most instructors would agree that the worst error you could make in this problem (the one that shows the least understanding) is to say the mass M exerts a force **F** on mass m.

In the case at hand, it is easiest to use Equation (A), because it has only one unknown.

$$F = (M + m)a$$
$$57\,N = (95\,kg)\,a$$

$$a = 0.60\,m/s^2$$

Substituting this result into Equation (C) gives

$$P = ma$$
$$= (30\,kg)\,(0.60\,m/s^2)$$
$$= 18\,N$$

The force on object m will always be smaller in magnitude than F. In some sense object M shields object m from the full effect of F.

* * * * * * * * * * * * * *

Example 4-7

A mass M is attached to a string of length R and swung in a vertical circle.

(a) Prove that the tension T_1 is greater than Mg at the bottom of the circle.

(b) What minimum velocity must the mass have at the top of the circle in order that the string does not go slack?

Solution: It is instructive to consider parts (a) and (b) of this problem at the same time. Below we present their solutions side by side.

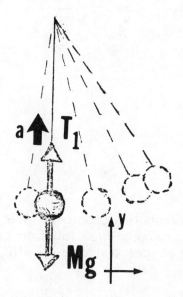

At the bottom

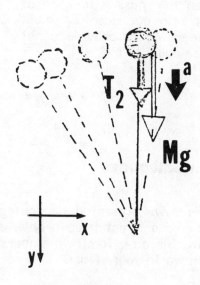

At the top

Because the mass M is moving in a circle, there is a centripital acceleration toward the center of the circle (upwards), and application of Newton's Second Law yields,

$$\sum F_y = T_1 - Mg = Ma$$
$$a = v_t^2/R$$
$$T_1 = Mg + Mv_t^2/R$$

where v_t is the tangential velocity at the bottom of the circle. We have proven that T_1 is greater than Mg (since we are adding a positive number to Mg).

* * * * * *

The most common error made by students in this type of problem is writing

$T_1 = Mg$

The above result is only true when the mass is hanging at rest. In our case T_1 must exceed Mg to provide a net force equal to Ma.

In this case take the +y direction to be downward, since both T_2 and Mg are downward (a string always pulls -- never pushes). Newton's Second Law gives:

$$\sum F_y = T_2 + Mg = Ma$$
$$a = v_t^2/R$$
$$T_2 = Mv_t^2/R - Mg$$

Study the previous equations. One can always choose v_t sufficiently large such that T_2 is greater than zero. Now let v_t start decreasing from this value. Eventually T_2 will equal zero, and for even smaller v_t the string will be slack.

Thus, asking for the minimum velocity of the mass at the top of the swing corresponds to solving the equation

$$T_2 = 0$$

which yields

$$v_t^2 = Rg$$
$$v_t = \sqrt{Rg}$$

* * * * * * * * * * * * * *

4.5 Problem Categories

In Chapter 3 we presented two categories of kinematics problems. We called them throw-up problems and catch-up problems. Force problems also fall into categories or families, but in this case family membership is not so cut and dried. Still, classifying problems is well worth your effort.

Next we give examples of two categories of force problems which often appear on examinations.

Category I -   Elevator Problems

Please note that the term "elevator" is a descriptive label to identify the category. Example 4-8 is the only example that really refers to an elevator.

* * * * * * * * * * * * * * *

Example 4-8

A baby of mass m sits on a scale in an elevator.

(a) If the elevator accelerates upward with an acceleration a, what weight does the scale read?

(b) If the elevator is moving downward and begins slowing down with deceleration a, what weight does the scale read?

(c) If the elevator cable breaks, and the elevator goes into free fall, what weight does the scale read?

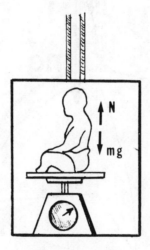

Figure 4.6

Solution: The trick in this problem is to realize that the scale reads the magnitude of **N**, <u>and not mg</u>.

We may now get the magnitude of **N** by applying Newton's Second Law <u>to the baby</u>. In parts (a) and (b) the acceleration is upward. This fact is obvious in part (a), but not at all obvious in part (b). In part

part (c) the acceleration is downward. Newton's Second Law yields

$$\sum F_y = N - mg = ma \quad (+y \text{ is upward})$$

$$N = mg + ma$$

(weight greater than mg)

for parts (a) and (b);

$$\sum F_y = mg - N = ma \quad (+y \text{ is downward})$$

$$N = mg - ma$$

for part (c). For free fall, we set $a = g$ in part (c) and obtain the result that $N = 0$. Thus an object in free fall is weightless.

* * * * * * * * * * * * * * *

Example 4-9

Does a person weigh more at the equator or the north pole?

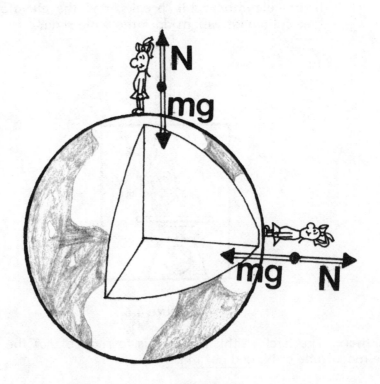

Figure 4.7

Solution: This problem is very similar to Example 4-8. In asking for the weight, we are asking what a scale would read if the person were standing on a scale. <u>So we are asking for the magnitude of the normal force.</u>

At the north pole the centripetal acceleration is zero because the person is standing at the axis of rotation, so Newton's Second Law says

$$N - mg = 0$$

$$N = mg$$

At the equator there is a centripetal acceleration directed toward the center of the earth. Taking this direction as "positive," Newton's Second Law gives

$$mg - N = ma$$

$$a = v_t^2/R$$

$$N = mg - mv_t^2/R$$

$$= mg\left(1 - \frac{v_t^2}{Rg}\right)$$

where v_t is the tangential speed of a point on the equator of the earth. Using 3960 mi for the mean radius of the earth we get

$$v_t = \frac{\text{circumference of earth}}{24 \text{ hours}}$$

$$= \frac{2\,\pi\,(3960 \text{ mi})}{24 \text{ hr}} \times \frac{5280 \text{ ft}}{\text{mi}} \times \frac{1 \text{ hr}}{3600 \text{ s}}$$

$$= 1521 \text{ ft/s}$$

Using this value of v_t, and g = 32.09 ft/s^2,

$$N = 0.9965\,mg,$$

So a person weighs slightly less at the equator.

* * * * * * * * * * * * * * *

Category II –   **Problems described by the diagram:**

56

Example 4-10

A small sphere of mass m is hung from the ceiling of an accelerating train by a string. The string makes an angle θ with the vertical. Show that tan θ = a/g.

Solution: Applying Newton's Second Law to Fig. 4.7 we get

x-direction:

$$T \sin \theta = ma$$

y-direction:

$$T \cos \theta = mg$$

Dividing the above equations gives the desired result:

$$\frac{\sin \theta}{\cos \theta} = \tan \theta = \frac{a}{g}$$

A common error is to have the mass swing in the wrong direction. You get the correct direction by imagining that the mass remains behind as the train jerks forward.

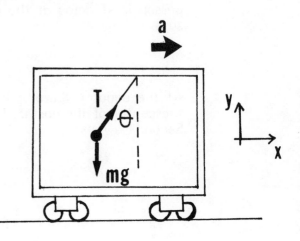

Figure 4.8

* * * * * * * * * * * * * *

Example 4-11

A circular road is banked at an angle θ so that an automobile can negotiate it at a speed v without the aid of friction. If the radius of curvature of the road is R, show that tan θ = v²/(Rg).

Solution: This example is almost identical to the previous one. Just replace the label T by N, and since the car is moving in a circle, a = v²/R. Newton's Second Law now yields the result:

x-direction:

$$N \sin \theta = mv^2/R$$

y-direction:

$$N \cos \theta = mg$$

$$\tan \theta = v^2/(Rg)$$

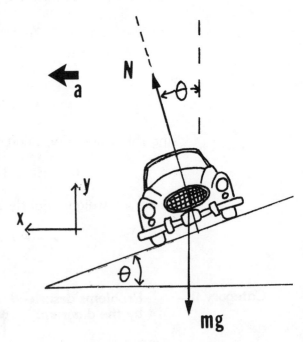

Figure 4.9

* * * * * * * * * * * * * *

In Figure 4.10 we show an arrangement which is called a conical pendulum. The mass m is moving in a circle of radius R with speed v. From the previous two examples you should immediately conclude that tan θ = v²/(Rg). Getting the tension involves some tricky algebra. See if you can obtain the result

$$T = m\sqrt{\frac{v^4}{R^2} + g^2}$$

Hint: $T^2 \sin^2\theta + T^2 \cos^2\theta = T^2$

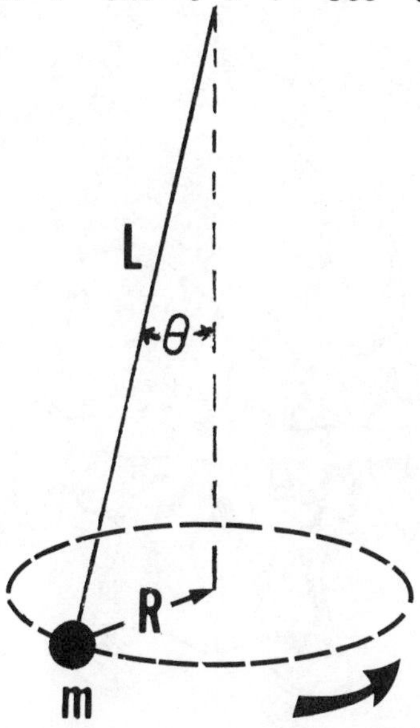

Figure 4.10 A mass m suspended from a string of length L swings so as to describe a circle. The string describes a right circular cone of semiangle θ.

5 WORK, ENERGY, POWER

肉体は上手に使わないと損をする？

Figure 5.1

5.1 Introduction

In Chapter 4 we discussed the concept of a force as defined by Newton through its effect <u>on</u> a body. The effect we emphasized was acceleration. You learned to associate <u>acceleration</u> with forces.

Sometimes we are aware of the presence of forces even though the body on which they are acting does not accelerate. For example, when we drive a car at <u>constant velocity</u>, the force on the wheels is balanced by frictional forces and air resistance; so the sum of the forces on the car is zero and there is no acceleration. Yet we must continue supplying fuel to the engine, and are aware of the existence of forces. Apparently there is something associated with forces other than acceleration. This "something else" is <u>work</u>. The contents of this chapter may be summarized by the statement,

<u>Forces do work and transform energy.</u>

In this context, both <u>work</u> and <u>energy</u> are technical terms of Physics, and their meanings are not to be confused with those in everyday usage.

5.2   Work

Suppose that you run up a flight of stairs, or lift weights, or swim a few lengths of a pool. What do each of these activities consist of from a scientific point of view? Simple: in each case you are exerting a force (to overcome gravity in climbing stairs or lifting weights, to overcome the drag of the water in swimming). But that is not all: <u>the force is being exerted through a distance.</u> You are climbing <u>up</u> the stairs; you are <u>lifting the weights</u>, or you are swimming <u>a few lengths</u>. In other words you are doing <u>work</u>.

Work by a Constant Force

In Physics the work done by a constant force **F** acting through a displacement **s** is defined to be (see Figure 5.2)

$$W = Fs \cos\theta \qquad (5.1\ a)$$

where θ is the angle between the vectors **F** and **s**. In terms of the scalar product of two vectors, work can be written as

$$W = \mathbf{F \cdot s} \qquad (5.1\ b)$$

So we see that work is a scalar quantity, <u>not a vector</u>. However, work <u>is</u> an algebraic quantity; it can be plus or minus.

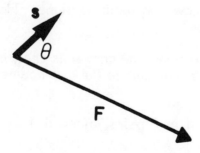

Figure 5.2 In order for work to be done, there must be a component of **s** in the direction of **F**.

Notice that <u>zero</u> work is done if the angle between the force **F** and the displacement **s** is 90º, since cos $\overline{90º}$ = 0. This is always the case in uniform circular motion. A circular path may be thought of as lots of little displacements **Δs** tangent to a circle.

Figure 5.3

We have already learned (in Chapter 4) that the force producing uniform circular motion points <u>inward</u> along the radius of a circle. Thus during any **Δ s** in Figure 5.3, the force is perpendicular to **Δ s**, and <u>no</u> <u>work</u> <u>is</u> <u>done</u>.

An Examination Favorite

A favorite short answer question (either a multiple choice question, or part of a larger question) is one of the following:

An object attached to a string is swung in a vertical circle. How much work is done by the tension in the string?

An object slides down an inclined plane. How much work is done by the normal force?

In each of the above questions, the answer is zero, because the force in question is always perpendicular to the displacement.

The meaning of <u>work</u> as used in Physics, often differs considerably from everyday usage.

* * * * * * * * * * * * * * *

Example 5-1

Exerting a force **mg,** a dockworker slowly lifts a box of mass m from the floor, to a height h, and then carries it through a horizontal displacement **s.** How much work is done by the dockworker?

(a)   **(b)**

Figure 5.4

Solution: The process described above divides naturally into two parts --

Part 1   In Figure 5.4 (a) the worker exerts a force mg upward as he lifts the box a distance h. The force is in the same direction as the displacement, so the angle θ appearing in Equation (5.1) is <u>zero</u> degrees. The work done in lifting the box is

$$W = mgh \cos \theta = mgh$$

Part 2   In Figure 5.4 (b) no work is done in transporting the box horizontally. The reason is that the force exerted by the worker is still mg upward, but the displacement is perpendicular to the force so,

$$W = (mg)(\text{horizontal displacement}) \cos 90^\circ = 0$$

Thus the total work done is mgh + 0 = mgh.

According to Equation (5.1), the work done by the force **F** is negative if $\theta = 180^\circ$. Remember that <u>negative work is a valid concept</u>. Suppose that, after transporting the box horizontally, the dockworker lowers the box onto the ground. In this process he does an amount of work mgh cos 180° = -mgh, and he has done total work

$$mgh \qquad + \; 0 \qquad + \; (-mgh) \qquad = \qquad 0$$

lifting	displacing	lowering	=	zero
it	it	it		total
	horizontally			work

The answer <u>zero total work</u> is expected because there has been <u>zero net displacement</u> in the vertical direction. But we required the concept of negative work to get this result.

* * * * * * * * * * * * * * *

A labor union would strongly dispute our definition of work in Example 5-1. In fact a laborer would consider it work just to stand still and hold a heavy object. <u>And the laborer would be correct</u> -- note the little fat man sweating in Figure 5.1. In this case work is being done because his arm muscles are under tension. There is lots of electrical activity in muscle fibers which are under tension. Chemical reactions occur, in which electrons are transferred from one site in an atom to another, under the influence of electrical forces. Hence work is being done; a force (electrical) acts through a distance (of the order of atomic distances). This work is converted to heat, and the little fat man in Figure 5.1 sweats.

If you think about it, the ability to do work is the ability to make things happen - to change things. Hence it is at the basis of all existence, all events, both biological and otherwise. For instance, consider thinking. Philosophical considerations aside, we know that in order to think, the brain must be undergoing many complex biological processes. In each of these processes chemical reactions occur, and as discussed above, work is done.

A system which has the ability to do work is said to posess <u>energy</u>. We shall next discuss energy of motion.

5.3 Kinetic Energy—The Work-energy Theorem

In Section 5.2 we discussed the work done by a single force; in this section we consider the work done by the net force. If a constant net force acts on a body and pushes it through a displacement $x - x_o$, Equation (5.1) tells us that

$$W = F^{net}(x - x_o) \tag{5.2}$$

But constant force means constant acceleration, and for constant acceleration, Equation (3.4) tells us that

$$x - x_o = \frac{v^2 - v_o^2}{2a}$$

(5.3)

Substituting Eq. (5.3) into Eq. (5.2), and using $F^{net} = ma$, gives

$$W = ma\left(\frac{v^2 - v_o^2}{2a}\right)$$

$$W = \frac{1}{2}mv^2 - \frac{1}{2}mv_o^2$$

(5.4)

Equation (5.4) is referred to as the <u>Work-energy Theorem</u>. The quantity ½ mv² is defined to be the <u>kinetic energy</u> or energy of motion. The traditional symbol for kinetic energy is K.

We have derived the Work-energy Theorem for the case of constant net force and displacement in a straight line. But one can prove (using calculus) that the Work-energy Theorem is valid even if the force is not constant, and for arbitrary displacement (along any curved path in three dimensional space). <u>The work done by the net force acting on a body during an arbitrary displacement is equal to the change in the kinetic energy.</u>

* * * * * * * * * * * * * *

Example 5-2

A 10 kg object moves along a complicated path in the x-y plane (see Figure 5.5). The object has an initial velocity of 10 m/s, and its final velocity is 12 m/s. How much work has been done by the net force?

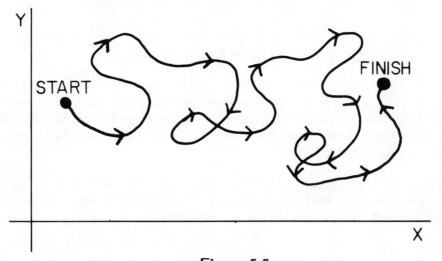

Figure 5.5

Solution: The crazy path is there to trick you. The work done is equal to the change in kinetic energy, and there is enough information given to get the initial and final kinetic energies. So the work done by the <u>net</u> <u>force</u> is

$$W = \frac{1}{2}mv_f^2 - \frac{1}{2}mv_i^2$$
$$= \frac{1}{2}(10\text{ kg})(12\text{ m/s})^2 - \frac{1}{2}(10\text{ kg})(10\text{ m/s})^2$$
$$= 220\text{ J}$$

* * * * * * * * * * * * * *

Example 5-3

A satellite is orbiting the earth in a circular path at constant speed. Calculate the amount of work done by the gravitational force.

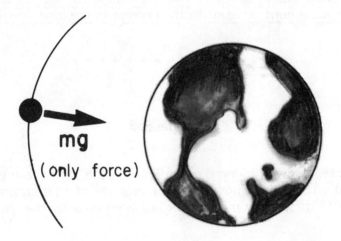

Figure 5.6

Solution: This is a variant of the question on page 60. You must realize that the only force acting on the satelite is the gravitational force (see Figure 5.6). Since the force is always perpendicular to the displacement, no work is done.

You could have arrived at the "zero work" answer in a simpler way. You were told that the speed remained constant, so there is no change in kinetic energy, and thus no work done!

* * * * * * * * * * * * * *

Example 5-4

A box of mass 25 kg has an initial speed of 20 m/s when x = 0. It slides on a rough surface. If the coefficient of kinetic friction between box and surface is 0.25, in what distance x does the box come to rest?

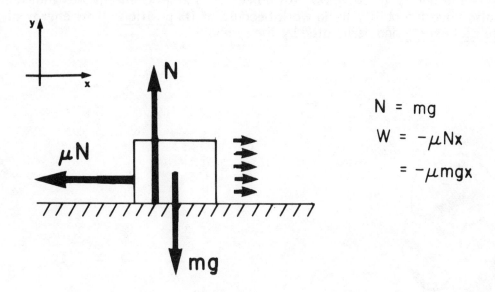

$$N = mg$$
$$W = -\mu Nx$$
$$ = -\mu mgx$$

Figure 5.7

Solution: The work energy theorem gives the answer in one step. In Fig. 5.7, the displacement and the frictional force are in opposite directions, so the work done by the frictional force is negative.

Work by friction	=	Final kinetic energy	−	Initial kinetic energy
$-\mu mgx$	=	0	−	$\frac{1}{2} mv^2$

$$\mu gx = \frac{1}{2} v^2$$
$$(0.25)(9.8 \text{ m/s}^2)\, x = \frac{1}{2}(20 \text{ m/s})^2$$
$$x = 81.6 \text{ m}$$

You can do the problem in two steps using $\Sigma F = ma$ to obtain a, and then using $v^2 = v_0^2 + 2ax$ to get x. But the beauty of the work-energy theorem is that you don't have to worry about forces (which are vectors) and equations of motion. You can get your answer by solving a single equation which involves only scalar quantities.

5.4 Potential Energy—Conservation of Energy

Energy is the ability to do work. An object with kinetic energy has that ability. An object may also have the ability to do work because of its <u>position</u>. This energy of position is called <u>potential energy</u> and designated by the symbol U.

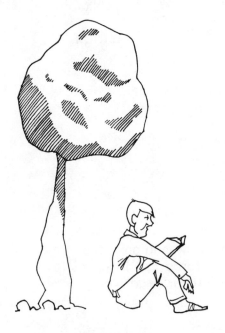

Figure 5.8

In Figure 5.8, the boulder starts out with pure (gravitational) potential energy which is continuously converted to kinetic energy as the boulder falls. Finally the boulder does work in pushing Herb into the ground.

There are other kinds of potential energy. In this chapter we shall also treat the energy of position of an object at the end of a spring. In this case $U = \frac{1}{2} kx^2$ where x is the distance that the spring has been stretched (or compressed) from equilibrium.

Figure 5.9

Conservation of Energy

If there are no frictional forces present, the three kinds of energy we have discussed so far (kinetic, gravitational potential, and elastic potential) satisfy a conservation principle:

$$\text{total mechanical energy} = \tfrac{1}{2}mv^2 + mgy + \tfrac{1}{2}kx^2 = \text{constant}$$

or

$$\textit{initial mechanical energy} = \textit{final mechanical energy}$$

$$\tfrac{1}{2}mv_i^2 + mgy_i + \tfrac{1}{2}kx_i^2 = \tfrac{1}{2}mv_f^2 + mgy_f + \tfrac{1}{2}kx_f^2 \tag{5.5}$$

We use Equation (5.5) in the following examples.

* * * * * * * * * * * * * *

Example 5-5

A woman is in a jet liner at an altitude of 2500 m, traveling horizontally at a velocity of 300 m/s. Attempting to enter the bathroom, she accidentally opens the rear door and steps out into space. Assuming no air resistance, find the magnitude and direction of her velocity just before she strikes the ground.

Solution: Aside from numerical changes, this example is the same as part (b) of Example 3-7. Conservation of energy is much easier to apply than the kinematics equations. Taking the ground as y = 0, we have

Kinetic plus potential   =   Just kinetic energy
energy at top   at bottom

$$\tfrac{1}{2} m (300 \text{ meters/s})^2 + m (9.8 \text{ meters/s}^2)(2500 \text{ meters}) = \tfrac{1}{2}mv^2$$

(divide out the m's before doing calculations)

$$v^2 = \left((300)^2 + 2(9.8)(2500) \right) \text{meters}^2/\text{s}^2$$

$$v = 373 \text{ m/s}$$

Recall that v_x never changes from its initial value of 300 m/s. To get the direction of the velocity at the ground we need v_y. From the Pythagorean theorem:

$$v_y^2 = v^2 - v_x^2 = (373 \text{ m/s})^2 - (300 \text{ m/s})^2$$

$$v_y = -221 \text{ m/s}$$

$$\tan \theta = 221/373 = 0.592$$

$$\theta = 30.6^\circ \text{ (into fourth quadrant)}$$

Note how <u>conservation of energy</u> simplifies this problem. You do not have to worry about the fact that the velocity is a vector in solving for v; it doesn't matter that $\mathbf{v_o}$ and $\mathbf{v}$ are in different directions.

* * * * * * * * * * * * * *

Example 5-6

A child's toy consists of a piece of plastic attached to a spring, with spring constant 400 N/m. The spring is compressed against the floor a distance of 3 cm, and the toy is released. If the mass of the toy is 80 g, what is its speed when it has risen to a height of 20 cm?

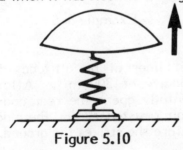

Figure 5.10

Solution: Once again <u>conservation of energy</u> lets us do this type of problem in a single step. We have purposely mixed units in this problem, so be sure to convert to S.I. units before beginning. It's sad when you "understand" everything, and lose points on an examination because you used grams rather than kilograms in an equation!

$$m = 80 \text{ gm} = 0.08 \text{ kg}$$
$$x_i = 3 \text{ cm} = 0.03 \text{ meters}$$
$$x_f = 20 \text{ cm} = 0.20 \text{ meters}$$

<table>
<tr><td align="center">elastic potential
energy</td><td align="center">kinetic
energy</td><td align="center">gravitational
potential energy</td></tr>
</table>

$$\tfrac{1}{2} (400) (0.03)\text{kg-m}^2/\text{s}^2 = \tfrac{1}{2} (0.08 \text{ kg})v^2 + (0.08) (9.8) (0.2) \text{ kg-m}^2/\text{s}^2$$

$$v = 0.215 \text{ m/s} = 21.5 \text{ cm/s}$$

* * * * * * * * * * * * * *

5.5 Frictional Effects—Nonconservative Forces

So far we have discussed the kinetic energy and the potential energy of an object in a gravitational field and at the end of a spring. The gravitational and spring forces have a special property -- they are <u>conservative</u> forces. That is, the work done when these forces act through a displacement, depends only on the starting and finishing positions, and not on the particular path taken. It is only for such forces that one can define a potential energy function.

Frictional forces are said to be nonconservative. The work done by such forces <u>does</u> depend on the path taken between two points. When frictional forces are present, total mechanical energy is no longer conserved; instead we have the result

<table>
<tr><td align="center">Initial kinetic
plus potential energy</td><td align="center">is greater than</td><td align="center">Final kinetic
plus potential energy</td></tr>
</table>

$$K_i + U_i > K_f + U_f \tag{5.6}$$

In this case the final mechanical energy is less than the initial mechanical energy. The reason is that the frictional forces have done work, and energy has been dissipated as heat (see Section 11.1). Using the work energy theorem we can modify Equation (5.6) and restore the equal sign

$$K_i + U_i = K_f + U_f + \left| \begin{array}{l} \textit{work done by} \\ \textit{frictional forces} \end{array} \right| \tag{5.7}$$

* * * * * * * * * * * * * * *

Example 5-7

A box of mass 2.5 kg slides down the inclined plane shown in Figure 5.11. The speed at the bottom is 5.0 m/s. What is the coefficient of friction between box and plane?

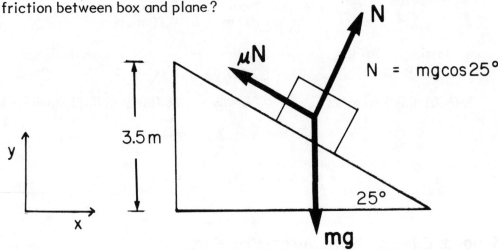

Figure 5.11

Solution: Take ground level as zero potential energy, and calculate the potential energy at the top of the plane, and the kinetic energy at the bottom.

$$U \text{ (top)} = mg \text{ (3.5 meters)} = (2.5 \text{ kg}) (9.8 \text{ meters/s}^2) (3.5 \text{ meters})$$
$$= 85.8 \text{ J}$$

$$K \text{ (bottom)} = \tfrac{1}{2} mv^2 = \tfrac{1}{2} (2.5 \text{ kg}) (5 \text{ meters/s})^2$$
$$= 31.3 \text{ J}$$

The kinetic energy at the bottom is less than the potential energy at the top, so work must have been done against the frictional force. The magnitude of the work done against the frictional force is $(\mu N) \times$(length of the plane). If we call the length of the plane L,

$$L = \frac{3.5 \text{ meters}}{\sin 25^\circ},$$

and we may write

$$U \text{ (top)} - K \text{ (bottom)} = \mu N L$$
$$= \mu \; mg \cos 25^\circ L$$
$$85.8 \text{ J} - 31.3 \text{ J} = \mu (2.5 \text{ kg}) (9.8 \text{ meters/s}^2) \cos 25^\circ \left(\frac{3.5 \text{ meters}}{\sin 25^\circ} \right)$$
$$54.5 = 183.9 \; \mu$$
$$\mu = 0.296$$

5.6   A Combination Problem

Professors love problems which test more than one chapter at a time; these often appear on examinations.   Below we give an example which tests energy conservation and centripetal acceleration.

* * * * * * * * * * * * * *

Example 5-8

A rollercoaster starts at rest a distance 2.8 R above the ground, and goes around a loop of radius R, as shown in Figure 5.12. Will the rollercoaster leave the track at Point P?

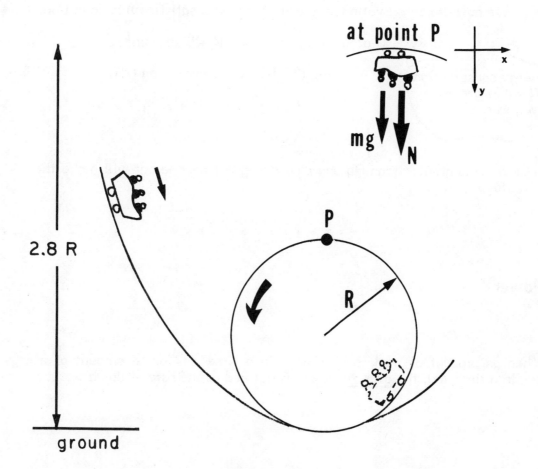

Figure 5.12

Solution: Let's concentrate on Point P. With N representing the (downward) normal force of the track, Newton's Second Law gives

$$\sum F_y = N + mg = ma$$

$$N + mg = m \frac{v^2}{R}$$

When the car leaves the track, it no longer feels the normal force exerted by the track; in other words, N = 0. Setting N = 0 in the previous equation gives the condition we desire. Since N cannot be directed upwards (which would give it a negative sign in our coordinate system),

$$\frac{mv^2}{R} > mg \quad \text{(when car remains on track)}$$

and

$$\frac{mv^2}{R} = mg \quad \text{(is the smallest value of } v^2 \text{ for which car makes the loop)}$$

Thus $\quad v^2 \geq Rg$ to stay on the track

We now use conservation of energy to see if v satisfies this condition.

For energy conservation, take U = 0 at ground and +y upward.

$$U \text{ at starting point} = K + U \text{ at point P}$$

$$mg(2.8R) = \tfrac{1}{2}mv^2 + mg(2R)$$

$$2.8\,Rg = \tfrac{1}{2}v^2 + 2\,Rg$$

$$v^2 = 1.6\,Rg$$

So v^2 is greater than Rg and everything is fine; the car <u>will</u> make the loop!

* * * * * * * * * * * * * * * *

5.7 Power

When an amount of work ΔW is done in a time Δt (or an amount of energy ΔE is expended in a time Δt), the <u>average power</u> is defined as the rate of doing work.

$$P_{av} = \frac{\Delta W}{\Delta t} \tag{5.8 a}$$

or

$$P_{av} = \frac{\Delta E}{\Delta t} \qquad (5.8\ b)$$

The instantaneous power P is found by considering smaller and smaller time intervals Δt, so

$$P = \frac{dW}{dt} \quad \left(or\ \frac{dE}{dt} \right) \qquad (5.9)$$

If we express dW in terms of the force F that is doing the work, we may rewrite Equation (5.9) in the useful form (see your textbook).

$$P = Fv\cos\theta \qquad (5.10)$$
$$= \textbf{F·V}$$

In Equation (5.10) **v** is the velocity of the particle, and θ is the angle between the vectors **F** and **v**.

Watt is the S.I. unit of power?

Watt is the S.I. unit of power.

1 watt = 1 joule/s

Remember that it is (power) × (time) that has units of energy. The electric company charges you for energy, so it charges you for kilowatt-hours, not for kilowatts.

In an introductory Physics course, most power problems (in mechanics) involve direct substitution in Equations (5.8) or (5.10). Usually there are no tricks in power problems. The professor just wants to see if you confuse power and energy.

Question: A 55 kg woman climbs up a vertical rope at a constant velocity of 1.2 meters/s. What average power must she supply?

Wrong answer: She is lifting her weight at constant speed, so F = mg in Equation (5.10)

$$
\begin{aligned}
P &= Fv \\
&= mgv \\
&= (55\ kg)\,(9.8\ m/s\ )\,(1.2\ m/s) \\
&= 647\ w
\end{aligned}
$$

Correct answer: The human body is approximately 20% efficient, so you must multiply the previous wrong answer by 5 to get the correct answer of 3235 w. In this case <u>we</u> get the dodo for being mean.

Question: Why does the electric company charge for kilowatt-hours rather than joules?

Answer: 1 kilowatt-hour × 1000 $\frac{\text{watts}}{\text{kilowatt}}$ × 3600 $\frac{\text{seconds}}{\text{hour}}$ × $\frac{1 \text{ joule}}{\text{watt-second}}$

= 3.6×10^6 joules.

People get angry enough being charged ~ 10¢ for a kilowatt-hour. Can you imagine their reaction if they received a bill claiming they had used <u>millions</u> of joules?

In an introductory Physics course, power plays a much smaller role than energy. However, in practical problems, both energy and power considerations can be important. The electric company may have the production capacity to supply your daily energy needs, but not your power needs. For example, lots of energy has to be supplied in a short time (between 2:00 and 5:00 PM) on a hot summer day when everyone turns on their air conditioners. Similarly, a weakened heart may be able to supply the daily energy needs of a person, but cannot handle sudden exertion. Sudden exertion means lots of energy being used in a short period of time, or large power production.

6 MOMENTUM, CENTER-OF-MASS

6.1 Introduction

In <u>America</u> you can depend on baseball, hot dogs, apple pie and Toyota. Conservation laws are even more dependable. Energy is conserved, not only in America, but everywhere in the <u>universe</u> -- the total energy in an isolated system never changes. From a philosophical point of view, conserved quantities provide a stability and continuity in our description of science. It is comforting to know that certain things never change, no matter how complicated the physical process involved. From a practical point of view, we have seen in Chapter 5, that the principle of <u>conservation of energy</u> enables us to solve various problems, in a way that is much simpler than alternative methods.

In Physics there is another, very useful conserved quantity called <u>momentum</u>. The momentum of a particle is defined as the product of its mass and its velocity, and is usually represented by the symbol **p**:

$$\mathbf{p} = m\mathbf{v} \tag{6.1}$$

Momentum is a vector quantity. For a system of many particles, the <u>total</u> <u>momentum</u> (represented by P) is defined to be the <u>vector sum</u> of the individual momenta.

$$\mathbf{P} = m_1\mathbf{v}_1 + m_2\mathbf{v}_2 + m_3\mathbf{v}_3 + \cdots \tag{6.2}$$

In the above formula, the first particle has mass m_1 and velocity $\mathbf{v}_1$, and so forth.

6.2 Conservation of Momentum

The principle of Conservation of Momentum states that <u>the total momentum of an isolated system never changes.</u>

Momentum conservation is a consequence of Newton's laws for forces. The forces acting on a system of particles can be divided into two classes:

1. The first class contains the forces which the different particles in a system exert on each other. These are called <u>internal forces</u>. According to Newton's Third Law, each such force has a partner which is equal in magnitude, and opposite in direction. <u>So the internal forces sum to zero.</u>

2. The second class consists of forces which are exerted from outside (external to) the system. For example, strings or sticks, or the gravitational force may reach in and exert forces on the particles in a system. These forces are called <u>external forces</u>.

If the sum of the <u>external forces</u> on a system is zero, the system is <u>isolated</u>, and <u>momentum is conserved</u>. This result is derived in all standard textbooks.

6.3 Applications of Momentum Conservation

Although momentum conservation holds only for isolated systems, in practice, it is OK to apply it to systems which are not <u>really</u> isolated. For example, during a <u>head-on</u> automobile collision, there are tremendous internal forces which the cars exert on each other along the line of the velocities. There are also external forces in this direction, such as the frictional forces between tires and the roadway. The effects of these external forces on the collision are negligible <u>compared</u> to those of the internal forces. So for practical pusposes we may assume that there are no external forces present, and that momentum is conserved.

There are three types of collisions --

1. <u>Elastic Collisions</u>: Both momentum and kinetic energy are conserved.

2. <u>Partially Inelastic Collisions</u>: Only momentum is conserved; kinetic energy is <u>not</u> conserved.

3. <u>Completely Inelastic Collisions</u>: Momentum is conserved; kinetic energy is not conserved, and furthermore, the objects stick together after the collision.

<u>Completely inelastic collisions are examination favorites</u>. Because the objects are in a single clump after the collision, the momentum conservation equations are simplified, and even collisions in two dimensions become fair game for a test.

Once again it is time for your mother (or your roomate) to enter the picture.

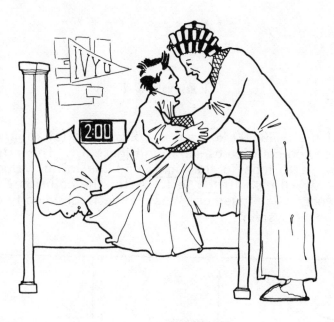

Figure 6.1

Your mother (or your roomate) must awaken you at 2:00 AM in the morning and shout,

"Inelastic Collision."

You must immediately respond,

"Momentum is conserved, but not kinetic energy."

Ten days of this procedure should train you properly to handle momentum problems on exams. An alternative approach is to memorize the following ditty, and repeat it each evening before going to sleep, and in the following morning upon awakening (again for ten days).

It's a well-established rule of thumb,

inelastic collisions conserve only momentum.

If you're hopeless and dumb as h _ _ _ ,

you'll conserve kinetic energy as well.

* * * * * * * * * * * * * * *

We now suppose that ten days have passed since you encountered the above material, and that you are ready for the following examples.

* * * * * * * * * * * * * * *

Example 6-1

A tank (of m_T = 40,000 kg) and a Volkswagen (of m_{VW} = 1000 kg) suffer a head on collision as shown in Figure 6.1. The initial velocity of the tank is 15 mph <u>to the right</u>, and the initial velocity of the Volkswagen is 60 mph <u>to the left.</u> After the collision the tank and the Volkswagen stick together. What is the final velocity of the tank and Volkswagen system?

Figure 6.2

Solution: The total momentum in the x-direction before the collision is

$$m_T v_T - m_{VW} v_{VW}$$

where v_T and v_{VW} are the speeds of the tank and the Volkswagen. After the collision the total momentum is

$$(m_T + m_{VW})V$$

where V is the desired answer.

<u>Conservation of momentum</u> says that momentum before equals momentum after --

$$m_T v_T - m_{VW} v_{VW} = (m_T + m_{VW})V$$

$$(40,000 \text{ kg}) (15 \text{ mi/h}) - (1,000 \text{ kg}) (60 \text{ mi/h}) = (41,000 \text{ kg})V$$

$$V = 13.2 \text{ mi/h}$$

(Note that the occupants of the tank were not affected very much by the collision!)

You must avoid two stumbling blocks in this problem.

1. Momentum is a vector (because it is proportional to the velocity which is a vector). Thus the momenta (velocities) of the tank and Volkswagen <u>must</u> <u>have</u> <u>opposite</u> <u>algebraic</u> <u>sign</u>. The major error in this problem is getting incorrect signs for the initial momentum.

$$m_T v_T + m_{VW} v_{VW}$$

2. We hope that you did not change miles per hour to meters per second! This procedure would waste lots of time (and possibly introduce errors) in an exam. It is OK to use "mongrel" units in the conservation of momentum equations. Mass is in mass units, and velocity is in velocity units; both are fundamental and don't affect each other.

* * * * * * * * * * * * * *

Example 6-2

Car A (m_A = 2000 kg) is moving east at 35 mph. Car B (m_B = 1500 kg) is moving north at 40 mph. The cars collide and stick together. What is the magnitude and direction of the final velocity of the two cars?

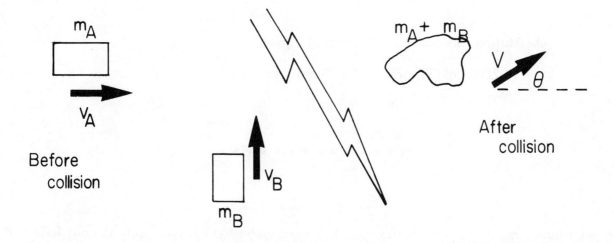

Figure 6.3

Solution: Use <u>conservation of momentum</u>, but remember that momentum is a vector, and must be conserved in both the east (x) direction and the north (y) direction.   The momentum before the collision is

$$P_x = m_A v_A$$

$$P_y = m_B v_B$$

The momentum after the collision is

$$P_x' = (m_A + m_B) V \cos \theta$$

$$P_y' = (m_A + m_B) V \sin \theta$$

<u>Conservation of momentum</u> yields the two equations:

(I) $m_A v_A = (m_A + m_B) V \cos \theta$ (for x direction)

(II) $m_B v_B = (m_A + m_B) V \sin \theta$ (for y direction)

The standard trick in solving for the unknowns (V and θ) is to divide Equation II by Equation I, obtaining

$$\frac{m_B v_B}{m_A v_A} = \tan \theta$$

$$\tan \theta = \frac{1500 \, (40)}{2000 \, (35)} = 0.857$$

$$\theta = 40.6^\circ$$

Substituting this result into Equation I gives

$$(2000 \text{ kg}) (35 \text{ m/s}) = (3500 \text{ kg}) V \cos 40.6^\circ$$

$$V = 26.3 \text{ m/s}$$

* * * * * * * * * * * * * * *

<u>Conservation of momentum</u> is often applied to very violent processes such as collisions and explosions. In such processes objects are ripped apart, and lots of heat is generated. For this reason they do not conserve kinetic energy. We have already discussed collisions in Examples 6-1 and 6-2. We consider an explosion in the next example.

* * * * * * * * * * * * * * *

Example 6-3

A steel container of mass M holds a small amount of nitroglycerine. It explodes into three fragments of equal mass. Two of the fragments have a speed of 35 m/s, and their velocity vectors make an angle of 60º. What is the magnitude and direction of the velocity of the third fragment?

Solution: You must realize that momentum is conserved, and that the initial momentum is <u>zero</u>. Thus the total momentum is <u>zero</u> after the collision. In Figure 6.3 we show the momentum vectors of the final three fragments; they must add to zero. For convenience we have chosen one of the momenta in the +x-direction.

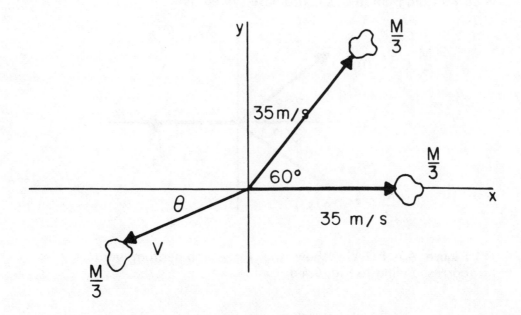

Figure 6.4 a

82

Momentum <u>conservation</u> says

$$\frac{M}{3}(35 \text{ m/s}) + \frac{M}{3}(35 \text{ m/s}) \cos 60^\circ - \frac{M}{3} V \cos \theta = 0 \qquad \text{(x-direction)}$$

$$\frac{M}{3}(35 \text{ m/s}) \sin 60^\circ - \frac{M}{3} V \sin \theta = 0 \qquad \text{(y-direction)}$$

Using the methods of example 6-2 we get

$$V \sin \theta = (35 \text{ m/s}) \sin 60^\circ$$

$$V \cos \theta = (35 \text{ m/s}) + (35 \text{ m/s}) \cos 60^\circ$$

$$\tan \theta = \frac{\sin \theta}{\cos \theta} = \frac{\sin 60^\circ}{1 + \cos 60^\circ} = 0.578$$

$$\theta = 30^\circ$$

$$V = (35 \text{ m/s}) \sin 60^\circ$$

$$= 60.6 \text{ m/s}$$

The previous approach works even when the masses and speeds are different. But there is an easier way to do <u>this</u> problem. Choose the coordinate system so that the fragment with unknown velocity moves in the positive x-direction. The other two fragments have equal mass and equal speed, so in some sense they must <u>behave</u> <u>symmetrically</u> (i.e., have similar behavior). Because the total momentum is zero, the correct diagram in this coordinate system is

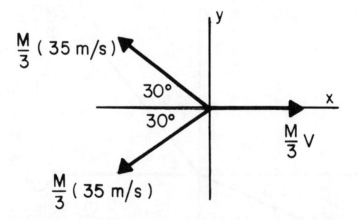

Figure 6.4 b We show the three momentum vectors corresponding to Fig. 6.4 a.

If the two angles in Figure 6.4 b were <u>not</u> 30°, there would be a <u>net</u> momentum in the y-direction. <u>This can't be</u>, since the total momentum is zero. So we already have the direction angles without doing any detailed mathematics!

The magic word is symmetry. We have used symmetry arguments to get the directions of the momenta vectors. Professors love symmetry arguments. In cases like this you will gain (rather than lose) Brownie points for avoiding detailed mathematics.

(Using conservation of momentum, show that you still obtain V = 60.6 m/s)

* * * * * * * * * * * * * * *

6.4   A Combination Problem—The Ballistic Pendulum

We give the ballistic pendulum problem a section of its own, because it is a most likely candidate for a mechanics examination. It is a test-maker's delight because it combines two important concepts -- <u>momentum</u> <u>conservation</u> and <u>energy</u> <u>conservation</u>. Furthermore, additional concepts can be included easily. For instance our version (Example 6-4) also tests the student's understanding of centripetal acceleration.

* * * * * * * * * * * * * * *

Example 6-4

A block of wood with mass 0.900 kg hangs suspended by a string. A bullet with mass 10.0 g and speed v smashes into it, burrows its way in (instantaneously), and the block swings up to a height h of 0.556 (Figure 6.5).

  (a) What was the initial velocity of the bullet?
  (b) What is the tension in the string at the bottom of the swing, <u>after</u> the bullet hits the block?

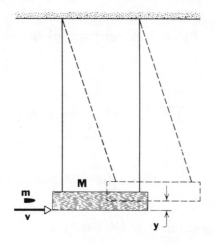

Figure 6.5

84

Solution: Part (a) of this problem divides naturally into two parts; what happens before and after the bullet comes to rest in the block.

Before:

We are dealing with an inelastic collision. Use <u>conservation of momentum</u> in this part. Initially the bullet has momentum mv, and the momentum of the block is zero. When the bullet has burrowed to rest, the block and bullet together have momentum (m + M)V. Thus

$$mv = (m + M)V$$

$$v = \left(\frac{m + M}{m}\right)V$$

After:

After the bullet has come to rest in the block, the block and bullet begin to rise. In this part of the problem, the sum of the potential and kinetic energy is conserved. <u>Conservation of energy</u> says

$$\tfrac{1}{2}(m + M)V^2 = (m + M)gh$$

$$V = \sqrt{2gh}$$

Combining the "before" and "after" results, we obtain

$$v = \left(\frac{m + M}{m}\right)\sqrt{2gh}$$

$$v = \frac{.910}{.010}\sqrt{2\,(9.8\ \text{m/s}^2)\,(0.0556\ \text{m})}$$

$$v = 300\ \text{m/s}$$

The tension T is <u>not</u> equal to (m + M)g.

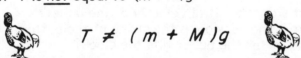

$$T \neq (m + M)g$$

Please review Example 4-7. The block is moving in a circle at the bottom of its path so there is a centripetal acceleration. Newton's Second Law gives

$$\sum F_y = T - (m + M)g = (m + M)\frac{v^2}{R}$$

$$R = \text{length of string}$$

$$T = (m + M)g + (m + M)\frac{v^2}{R}$$

As you can see, T is greater than (M + m)g.

* * * * * * * * * * * * * *

6.5   Center of Mass

The same reasoning that leads to the principle of <u>Conservation of Momentum</u> for an isolated system, tells us that the velocity of the center of mass of an isolated system never changes. In Figure 6.6 we show two objects with center of mass coordinates x_1 and x_2, and masses m_1 and m_2.

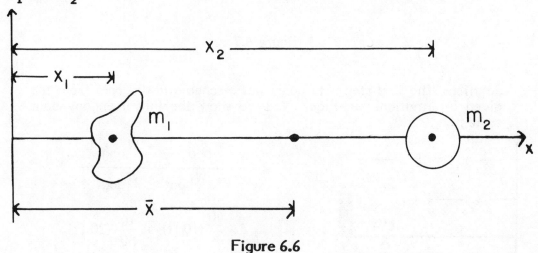

Figure 6.6

Recall (from your textbook) that the overall center of mass coordinate $\bar{x}$ is

$$\bar{x} = \frac{m_1 x_1 + m_2 x_2}{m_1 + m_2} \quad or \quad \frac{w_1 x_1 + w_2 x_2}{w_1 + w_2} \tag{6.3}$$

If no net <u>external</u> forces act on the two objects in Figure (6.6), and they are initially at rest, the center of mass will remain at rest. Certain problems are easier to solve using this idea, than by working with momentum. In particular, let's consider a simplified version of the famous "Snoopy" problem in Halliday and Resnick.

* * * * * * * * * * * * * * *

Example 6-5

A dog, weighing 10 lb, is standing at one end of a flatboat that is 20 ft long. He walks to the other end of the boat, and then halts. The boat weighs 40 lb, and one can assume there is no friction between it and the water. How far does the boat move?

86

Figure 6.7

Solution: The first step is to construct a schematic diagram from the given information. In Figure 6.8 we show the initial Snoopy-boat system.

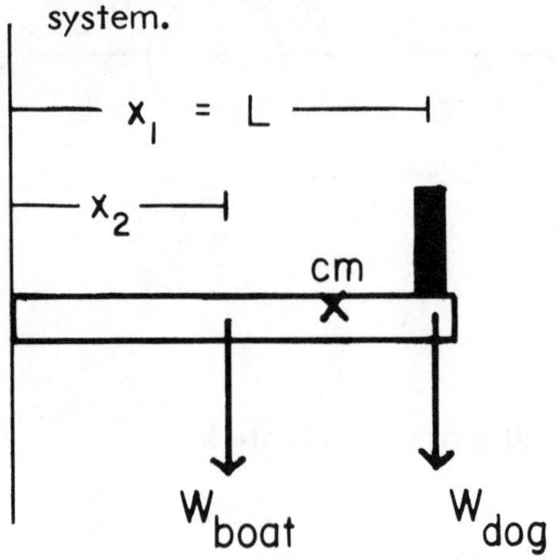

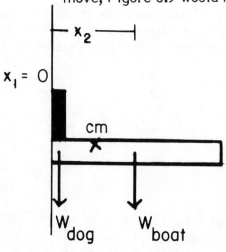

$$\bar{x} = \frac{W_{boat} \frac{L}{2} + W_{dog} L}{W_{boat} + W_{dog}}$$

$$= \frac{40}{50}(10 \text{ ft}) + \frac{10}{50}(20 \text{ ft})$$

$$= 8 \text{ ft} + 4 \text{ ft}$$

$$= 12 \text{ ft}$$

Figure 6.8

If Snoopy had walked to the end of the boat, and the boat did not move, Figure 6.9 would now describe the Snoopy-boat system.

$$\bar{x} = \frac{W_{boat} \frac{L}{2} + W_{dog}(0)}{W_{boat} + W_{dog}}$$

$$\bar{x} = 8 \text{ ft}$$

Figure 6.9

There are no external forces in the x-direction because there is no friction between the boat and the water. Thus the center of mass cannot move in this direction. Figure 6.9 must be wrong because the center of the mass does not move. To make it correct we must shift the center of mass to its original position. So the boat shifts 4 ft to the right.

* * * * * * * * * * * * * *

6.6 An Application of the Scalar Product

A frustrating aspect of many introductory Physics courses is that advanced ideas are often introduced, but not applied to realistic problems. The scalar product is an example of such an idea. It would be nice to see applications outside the chapter on vectors. To relieve your frustrations, we present below a "neat" application of the scalar product to <u>elastic</u> collisions.

* * * * * * * * * * * * *

Example 6-6

Paul the pool shark shoots ball A at ball B which is initially at rest. After they collide the balls go off at an angle θ with respect to one another. If the balls are point masses, and we can ignore their rotation and friction, what is θ?

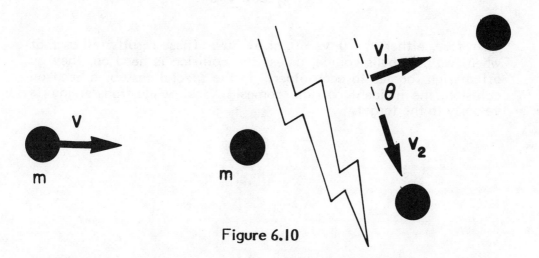

Figure 6.10

Solution: This is an elastic collision, so both momentum and kinetic energy are conserved. Using the labeling of Figure 6.10, we get

$$m\,\mathbf{v} \;=\; m\,\mathbf{v_1} \;+\; m\,\mathbf{v_2} \qquad\qquad \text{(conservation of momentum)}$$

$$\tfrac{1}{2}\,m\,v^2 \;=\; \tfrac{1}{2}\,m\,v_1^2 \;+\; \tfrac{1}{2}\,m\,v_2^2 \qquad\qquad \text{(conservation of energy)}$$

Dividing out the m's and the ½ m's, we obtain our working equations.

(A) $\mathbf{v} = \mathbf{v_1} + \mathbf{v_2}$

(B) $v^2 = v_1^2 + v_2^2$

Substitute Equation (A) into Equation (B) obtaining

(C) $(\mathbf{v_1} + \mathbf{v_2})^2 = v_1^2 + v_2^2$

STOP AND THINK! Remember that all the velocities are vectors. Thus their squares may be expressed in terms of dot products.

$$v_1^2 = \mathbf{v_1} \cdot \mathbf{v_1}$$

$$v_2^2 = \mathbf{v_2} \cdot \mathbf{v_2}$$

$$(\mathbf{v_1} + \mathbf{v_2})^2 = \mathbf{v_1} \cdot \mathbf{v_1} + \mathbf{v_2} \cdot \mathbf{v_2} + 2\,\mathbf{v_1} \cdot \mathbf{v_2}$$

Plugging these results into Equation (C) gives

$$v_1^2 + v_2^2 + 2\,\mathbf{v_1} \cdot \mathbf{v_2} = v_1^2 + v_2^2$$

$$2\,\mathbf{v_1} \cdot \mathbf{v_2} = 0$$

$$2\,v_1\,v_2 \cos\theta = 0$$

Therefore, either $v_1 = 0$, $v_2 = 0$, or $\theta = 90°$. These results tell us that when two pool balls collide, unless the collision is head on, they go off at right angles to each other. In the special case of a head-on collision, the incident particle comes to rest while transferring its velocity to the target.

7 TORQUE

7.1 Introduction

Up to this point you have dealt only with point masses. You have learned that a particle is in equilibrium when the sum of the forces acting upon it is zero.

But most objects in nature are <u>extended</u> objects; they occupy space. Consider the stick in Figure 7.1 -- it is an extended <u>object</u>. Although the sum of the forces F and –F is zero, from experience you know that the stick will rotate rather than remain at rest.

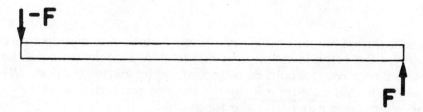

Figure 7.1

Rotations are caused by physical quantities called <u>torques</u>. From now on, not only is the magnitude and direction of the force exerted on an object important, but <u>WHERE</u> it is applied is crucial. For instance, you would not push at the hinge of a door to open it. A thin kid on a seesaw can balance his fat friend by moving farther out (see Figure 7.2).

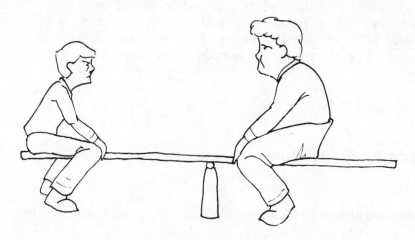

Figure 7.2

The torque is defined as the product of the force and the moment (or lever) arm. The moment arm is the perpendicular distance from the axis of rotation to the line along which the force acts. This definition becomes clearer if you examine Figure 7.3.

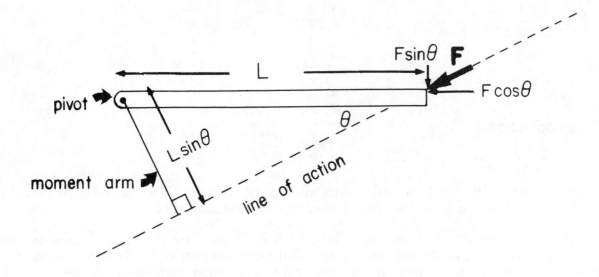

Figure 7.3

For your needs, torque comes in two varieties -- clockwise and counterclockwise. For example, in Figure 7.3 the force is making the rod turn in a clockwise direction. If the force were in the opposite direction, it would cause counter-clockwise rotation. We usually adopt the standard convention that counter-clockwise torques are positive, and clockwise torques are negative. For example, in Figure 7.3 the torque is

$$\tau = -F(L\sin\theta) \qquad (clockwise) \tag{7.1}$$

Torque is actually a vector which is made out of the force vector and a displacement vector by a moderately complicated multiplication called the cross product. Halliday and Resnick, for example, define torque this way. In an introductory Physics course you will almost always deal with rotation in a plane. For this case you need not worry about the full complexity of the cross product, because torques will always be clockwise or counter-clockwise.

7.2 Conditions for Equilibrium

For an extended object to be in equilibrium, two conditions must be satisfied.

I: Sum of $\sum F_x = 0$
   Forces = 0
              $\sum F_y = 0$

(7.2)

II: Sum of
    Torques = 0 $\sum \tau = 0$

We shall give you two tips for solving equilibrium problems involving torques, and immediately follow with illustrative examples.

1. Resolve all forces into vector components, and consider the torques produced by the individual components. For example, in Figure 7.3 the horizontal and vertical components of F (and corresponding lever arms) are

$$F_x = F\cos\theta, \text{ lever arm} = 0 \tag{7.3}$$

$$F_y = F\sin\theta, \text{ lever arm} = L \tag{7.4}$$

STOP! Equation (7.3) is crucial -- if you do not understand it, ask your professor or a friend to explain it. From Equation (7.3) we see that the horizontal component of F produces zero torque (because the lever arm is zero), while Equation (7.4) tells us that the vertical component produces a torque

$$\tau = -(F\sin\theta)L \quad (clockwise)$$

Naturally we obtain the same results as in Equation (7.2), except that the parentheses appear in different places. From the earlier discussion that led to Equation (7.2), it is obvious that you don't have to resolve your forces into components, for the purpose of calculating torques in equilibrium problems. However, you had better do so, or you will have lots of trouble with such problems.

2. When an object is in equilibrium, the sum of the torques is <u>zero about any axis</u>. So you have a choice. A good choice can make a problem simple; a bad choice can make a problem almost impossible to solve! Usually the good choice is an axis through which pass as many unknown forces as possible. These produce no torque, since their lever arms are zero, and you have reduced the number of unknowns in the torque equation (for equilibrium). This idea will become more clear after you have studied Example 7-1.

Beam problems and ladder problems are the most likely candidates for equilibrium problems on examinations. Follow the above rules and you will handle them easily.

* * * * * * * * * * * * * * *

Example 7-1

The strut in Figures 7.4 (a) and 7.4 (b) weighs 200 N and its center of mass is at its center. In both cases (a) and (b) find

(i)   the tension in the cable and

(ii) the horizontal and vertical components of the force exerted on the strut at the wall.

Solution: We shall treat cases (a) and (b) in parallel. The only difference is that it is more difficult to determine the lever arms in case (b).

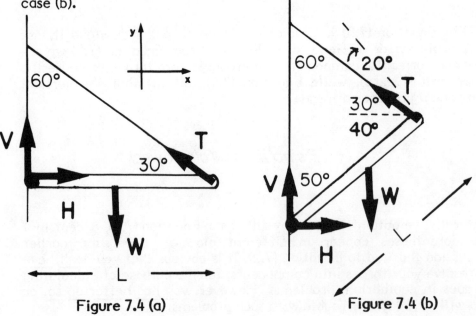

Figure 7.4 (a)   Figure 7.4 (b)

First draw all forces acting ON the beam. The string pulls (along its length) with a force T. The weight W acts straight down. These two forces are easy.

In accordance with our approach of always working with components, we have represented the force exerted by the pivot by its (unknown) horizontal and vertical vector components H and **V**.

We take the axis of rotation at the pivot. Thus H and **V** do not appear in the torque equation, allowing us to solve directly for T. The equations of equilibrium are

$$\sum F_x = H - T\cos 30^\circ = 0$$

$$\sum F_y = T\cos 60^\circ + V - W = 0$$

$$\sum_{\substack{counter- \\ clockwise}} \tau = T\cos 60^\circ (L) - W\left(\frac{L}{2}\right) = 0$$

Solving the torque equation for T, and substituting back in the force equation gives

$$T = \frac{\frac{W}{2}}{\sin 30^\circ}$$

$$T = 200 \text{ N}$$

$$H = (200 \text{ N})\cos 30^\circ$$

$$= 173 \text{ N}$$

$$V = W - T\sin 30^\circ$$

$$= 100 \text{ N}$$

First draw all forces acting ON the beam. The string pulls (along its length) with a force T. The weight W acts straight down. These two forces are easy.

In accordance with our approach of always working with components, we have represented the force exerted by the pivot by its (unknown) horizontal and vertical vector components H and **V**.

We take the axis of rotation at the pivot. Thus H and **V** do not appear in the torque equation, allowing us to solve directly for T. The equations of equilibrium are

$$\sum F_x = H - T\cos 30^\circ = 0$$

$$\sum F_y = T\cos 60^\circ + V - W = 0$$

$$\sum_{\substack{counter- \\ clockwise}} \tau = (T\cos 20^\circ) L - W\left(\frac{L}{2}\right)\sin 50^\circ = 0$$

Solving the torque equation for T, and substituting back in the force equation gives

$$T = \frac{\frac{W}{2}\sin 50^\circ}{\cos 20^\circ}$$

$$T = 81.5 \text{ N}$$

$$H = (81.5 \text{ N})\cos 30^\circ$$

$$= 70.6 \text{ N}$$

$$V = W - T\sin 30^\circ$$

$$= 159.25 \text{ N}$$

* * * * * * * * * * * * * *

Example 7-2

In Figure 7.5 a ladder 8.5 m long, of weight 350 N leans against a vertical, frictionless wall. The ladder is at rest, making an angle of 60 with the horizontal. Find the magnitude and directions of the forces R and P, where P is the force exerted on the ladder by the wall, and R is the force exerted on the ladder by the ground.

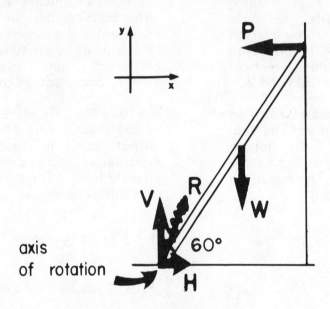

Figure 7.5

Solution: This is the infamous ladder problem. In our opinion, more students have floundered on the shoals of this problem, than on any other problems in Physics.

In Figure 7.5 we have already performed the first step in an equilibrium problem by breaking up the forces into horizontal and vertical vector components. The next step is to write down the force and torque equations, taking the axis of rotation as in Fig. (7.5).

$$\sum F_x = H - P = 0$$

$$\sum F_y = V - W = 0$$

$$\sum_{\substack{counter- \\ clockwise}} \tau = P\,(8.5\text{ m})\sin 60^\circ - W\,\frac{(8.5\text{ m})\sin 30^\circ}{2} = 0$$

P is the only unknown in the torque equation -- solving this equation for P and substituting the results into the force equations gives us the answers: P = 101 N, V = 350 N, H = 101 N.

The magnitude and direction of R are

$$R = \sqrt{V^2 + H^2} = 364 \, N$$

$$\tan \theta = \frac{V}{H}; \quad \theta = 73.9$$

It seems easy doesn't it? We really do not understand why so many students go wrong on this problem. Being systematic and exerting a certain amount of thought are the secrets. For example we have seen students put H in the wrong direction on their force diagram. It is not the most serious error in the world, but a little sad, because common sense says otherwise. The ground doesn't help the ladder stay up by pulling to the left! (See Figure 7.6.)

Figure 7.6

It helps if you always think in terms of the forces acting <u>on</u> the ladder. The vertical wall exerts a force P to the left, because we are told that the wall is frictionless. A frictionless surface only exerts a force perpendicular to itself; never along itself.

* * * * * * * * * * * * * *

8 ROTATIONAL MOTION AND TORQUES

8.1   Introduction

In Chapter 7 we introduced torques and used them to describe extended objects in equilibrium. In that case the sum of the torques was zero. In this chapter we consider what happens when the sum of the torques is not zero; in this case the object rotates faster and faster about its center mass. It is generally agreed that the subject material of this chapter is the most difficult in an introductory Physics course.

When an object is not in rotational equilibrium,

$$\sum \tau = I\alpha \tag{8.1}$$

about the center of mass or a stationary point. In Equation (8.1) I is the <u>moment of inertia</u> and α is the <u>angular acceleration</u>. For you, these are two new concepts which you must master in order to handle rotational motion. In addition you will deal with <u>rotational kinetic energy</u> and <u>angular momentum</u> in this chapter.

8.2   Moment of Inertia

In rotational motion, the moment of inertia plays the same role that mass plays in straight line motion. It takes a large force to give a large mass a large acceleration in a straight line. Likewise, if an object has a large moment of inertia, it will take a large torque to give it a large angular acceleration. The moment of inertia depends on the mass of an object <u>and</u> the distribution of mass about the axis of rotation.

For a point mass m, the moment of inertia is

$$I = mr^2 \tag{8.2}$$

where r is the distance from the axis of rotation to the mass. For a group of masses m_1, m_2, m_3 ... (which are part of a rigid system)

$$I = m_1 r_1^2 + m_2 r_2^2 + m_3 r_3^2 + \ldots$$

$$= \sum_i m_i r_i^2$$

(8.3)

At this point you must be wondering why we keep talking about <u>point</u> masses. What's the <u>point</u>? In real life we don't seem to deal with groups of <u>point</u> masses; rather we deal with trees, books, chairs, tables and so forth.

The <u>point</u> is that most of the formulas that you have learned deal with <u>point</u> masses. Fortunately, a realworld object can always be divided (in your imagination) into very small parts, and we can apply our formulas for <u>points</u> to these small parts.

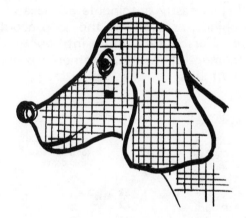

Figure 8.1 Rigid Dog

Evaluation of Equation (8.3) for the dog in Figure 8.1 isn't an easy job. One must do the evaluation numerically on a computer. The only simple case that we know, in which Equation (8.3) can be evaluated without a computer, or the use of calculus, is that of a hoop about an axis perpendicular to the plane of the hoop, and passing through its center (see Figure 8.2)

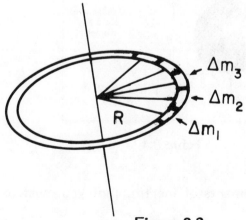

Figure 8.2

For this case we use Equation (8.3) in the following way:

$$I = \Delta m_1 R^2 + \Delta m_2 R^2 + \Delta m_3 R^2 + \ldots$$

$$= R^2(\Delta m_1 + \Delta m_2 + \Delta m_3 + \ldots)$$

$$= MR^2 \ (because \ \Delta m_1 + \Delta m_2 + \ldots = M) \tag{8.4}$$

The final result [Eq. (8.4)] is easily obtainable because R is the same for each Δ m. This is not true for any other object, or even for the hoop about some other axis. For all other cases one must integrate. Fortunately, in an introductory Physics course one rarely must calculate a complicated moment of inertia. Rather you are asked to use moments of inertia which are obtained from a table.

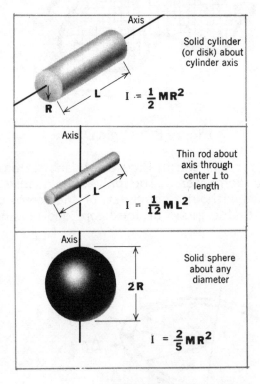

Table 8.1

There are two rules concerning moments of inertia, that you must be able to use (usually together with Table 8.1):

I. Moments of Inertia Add

If two bodies are rigidly connected, the total moment of inertia is the sum of the two individual moments of inertia.

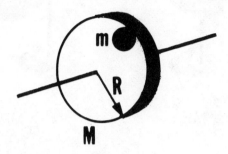

Figure 8.3   The mass m is a point mass at the edge of the disc.

As an example consider Figure 8.3. A uniform disc of mass M and radius R has a small mass m attached to it a distance r from the center (see Figure 8.2). The moment of inertia of the disc and point mass about an axis which is perpendicular to the plane of the disc, and passes through its center is

$$I = I\,(\text{disc}) + I\,(\text{point mass})$$

$$= \tfrac{1}{2}\,MR^2 + mR^2$$

2. Parallel Axis Theorem

Suppose the moment of inertia of an object of mass m about any axis through its center of mass is I_{cm}. The moment of inertia about any axis parallel to this axis, and a distance b away, is

$$I = I_{cm} + mb^2 \qquad (\textit{parallel axis theorem}) \tag{8.5}$$

We demonstrate the use of the parallel axis theorem in Figure 8.4.

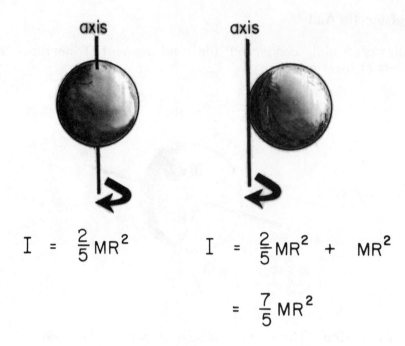

$$I = \frac{2}{5}MR^2 \qquad I = \frac{2}{5}MR^2 + MR^2$$

$$= \frac{7}{5}MR^2$$

Figure 8.4

8.3 Angular Velocity and Angular Acceleration

In Figure 8.5 we show a point P moving around a circle. The length s and the angle θ are changing with time. Starting with the definition of the angle in radians

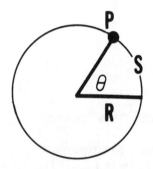

Figure 8.5

$$\theta = \frac{S}{R} \quad (\text{in radians}) \quad (8.6)$$

your textbook derives relations between the angular velocity ω and the tangential velocity, and between the angular acceleration α and the tangential acceleration.

$$\omega = \frac{v_T}{R} \tag{8.7}$$

$$\alpha = \frac{a_T}{R} \tag{8.8}$$

The angular velocity ω measures the rate at which the angle θ is changing in radians/second; α measures the rate at which ω is changing in radians/sec^2. The relations in Equations (8.7) and (8.8) <u>must</u> <u>be</u> <u>memorized</u>. They are used often in rotational motion problems, and you must have them at the "tip of your tongue" if you are going to do well on exams.

Like torque, angular velocity and angular acceleration are really vectors, but we shall not treat them as such. Since we shall only deal with motion in a plane, angular velocity (and angular acceleration) can be described as either clockwise or counterclockwise. So, for example, counterclockwise ω's are taken to be positive and clockwise ω's as negative, and for our purpose, angular velocities add like numbers, rather than vectors.

8.4   Angular Momentum

For each physical quantity in a straight line motion, there is a matching rotational one. Up to now we have considered the corresponding quantities

$$v \longleftrightarrow \omega$$
$$a \longleftrightarrow \alpha$$
$$m \longleftrightarrow I$$

and the corresponding equations

$$\sum F = ma \longleftrightarrow \sum \tau = I\alpha$$

Just as one develops energy and momentum conservation principles from $F = ma$ (for straight line motion) one can define the quantities

$$I\omega \quad (\text{angular momentum}) \tag{8.9}$$

$$\tfrac{1}{2}I\omega^2 \quad (\text{rotational kinetic energy}) \tag{8.10}$$

and develop associated conservation laws. In this section we deal with angular momentum. We treat angular momentum problems first, because they're the easiest.

Conservation of Angular Momentum

The angular momentum of an isolated system (no external torques) never changes.

Example 8-1

A famous cat juggler stands on a freely rotating platform, holding two screeching cats at arms lengths. The terrified tabbies jump onto his head. What happens to his angular velocity (see Figure 8.6)?

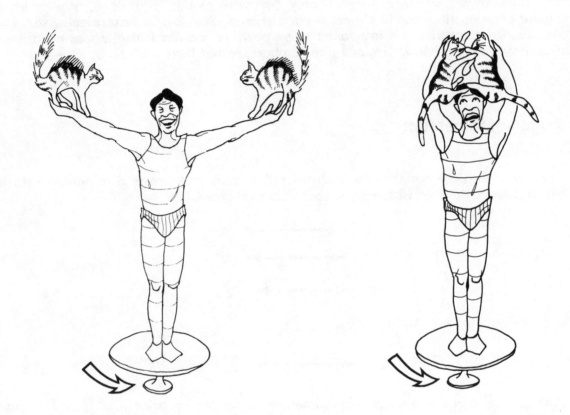

Figure 8.6

Solution: The cat juggler and cats form an isolated system so angular momentum is conserved.

$$I_a \omega_a = I_b \omega_b$$

$$\omega_b = \frac{I_a}{I_b} \omega_a$$

We have used the subscript a to represent Fig. 8.6 (a), and the subscript b to represent Fig. 8.6 (b). In case (a) there is more mass distributed farther from the axis than in case (b), so I_a is greater than I_b. Thus we have proven that the final angular velocity is greater than the initial angular velocity.

* * * * * * * * * * * * * * *

Example 8-2

A uniform disc with mass 1.5 kg and radius 0.25 m rotates with angular velocity 12 rad/s about a vertical, frictionless axis. A second uniform disc with mass 2.0 kg and radius 0.12 m, initially not rotating, drops onto the first cylinder (see Figure 8.7). The cylinders stick together.

(a)   Calculate the new angular velocity.

(b)   Show that energy is lost in this "collision".

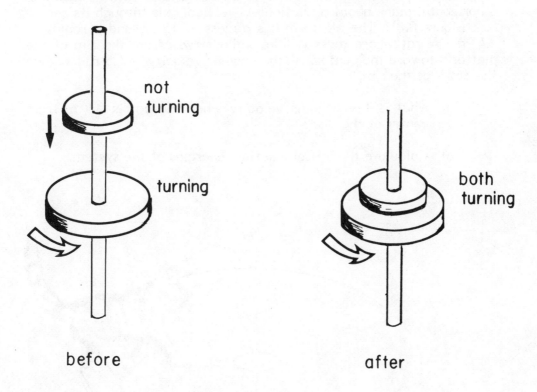

not turning

turning

both turning

before

after

Figure 8.7

Solution: (a) This is an inelastic collision! Angular momentum is conserved, but not kinetic energy. To get the final angular velocity we use conservation of angular momentum.

$$I_1 \omega_1 = I_2 \omega_2$$

$$I_1 = \tfrac{1}{2}(1.5 \text{ kg})(0.25 \text{m})^2 = 0.047 \text{ kg} \cdot \text{m}^2$$

$$I_2 = I_1 + \tfrac{1}{2}(2.0 \text{ kg})(0.12 \text{ m})^2 = 0.061 \text{ kg} \cdot \text{m}^2$$

$$\omega = \frac{I_1}{I_2}(12 \text{ rad/s}) = 9.3 \text{ rad/s}$$

(b) We calculate the initial and final kinetic energy.

$$K(\text{initial}) = \tfrac{1}{2} I_1 \omega_1^2 = 3.38 \text{ J}$$

$$K(\text{final}) = \tfrac{1}{2} I_2 \omega_2^2 = 2.64 \text{ J}$$

* * * * * * * * * * * * * * *

Example 8.3

A horizontal platform in the shape of a circular disc rotates freely in a horizontal plane about a frictionless vertical axis through its center (see Figure 8.8). The platform has a mass of 15 kg and a radius of 2.4 m. A cat whose mass is 5 kg walks slowly from the rim of the platform toward the center. If the angular velocity is 1.5 rad/s when the cat is at the rim,

(a) what is the angular velocity when the cat is 0.5 m from center of the disc?

(b) Calculate the initial and final energies of the system.

Figure 8.8

Solution: (a) The first step is to get an expression for the moment of inertia of the system. The cat is small compared to the disc, so we will treat it as a point mass. Since we have an isolated system (disc plus cat), we may use conservation of angular momentum to solve the problem.

$$I_1 \omega_1 = I_2 \omega_2$$

$$I_1 = \tfrac{1}{2}(15 \text{ kg})(2.4 \text{ m})^2 + (5 \text{ kg})(2.4 \text{ m})^2 = 72.0 \text{ kg} \cdot \text{m}^2$$

$$I_2 = \tfrac{1}{2}(15 \text{ kg})(2.4 \text{ m})^2 + (5 \text{ kg})(0.5 \text{ m})^2 = 44.5 \text{ kg} \cdot \text{m}^2$$

$$\omega_2 = \frac{I_1}{I_2} \ (1.5 \text{ rad/s}) = 2.43 \text{ rad/s}$$

(b)   We calculate the initial and final kinetic energies using Equation (8.10).

$$K(\text{initial}) = \tfrac{1}{2} I_1 \omega_1^2 = 81 \text{ J}$$

$$K(\text{final}) = \tfrac{1}{2} I_2 \omega_2^2 = 131 \text{ J}$$

This is an interesting result. In the previous example, we saw that angular momentum was conserved, while kinetic energy decreased. Here we have angular momentum being conserved, but kinetic energy increasing. Where did the energy come from? The answer is that in walking inward, the cat had to exert muscular effort and do work. So the extra kinetic energy of the system came from the Purina Cat Chow eaten earlier in the day!

* * * * * * * * * * * * * * *

8.5   Rotational Kinetic Energy

Suppose that an object is performing pure rotation about a stationary axis. (Pure rotation means that every point on the object has the same angular velocity -- not tangential velocity (!!!) -- about the axis.) For this case the kinetic energy may be written

$$K = \tfrac{1}{2} I \omega^2 \tag{8.11}$$

where I is the moment of inertia about the stationary axis. The term $\tfrac{1}{2} I \omega^2$ must now be included in the total mechanical energy along with the quantities

$$mgy, \ \tfrac{1}{2}mv^2, \ \tfrac{1}{2}kx^2$$

considered in Chapter 5.

* * * * * * * * * * * * * *

Example 8-4

A uniform rod of length L and mass M is free to rotate about a frictionless pivot at one end (see Figure 8.9). The rod is released from rest in the horizontal position.

    (a) What is the angular velocity of the rod when it is at its lowest position?

    (b) Determine the linear velocity of the <u>center of mass</u> when the rod is in the vertical position.

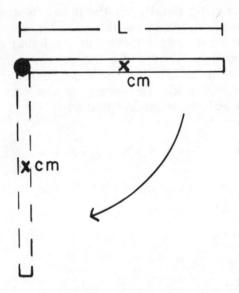

Figure 8.9

Solution: (a) The main trick here is to realize that the change in potential energy is (mg) × (the distance that the c.m. falls). We take the potential energy to be zero when the rod is horizontal; the kinetic energy is also zero in that position. In the vertical position, the rod has rotational kinetic energy and (negative) potential energy. Conservation of energy gives the result

$$0 = -Mg\frac{L}{2} + \frac{1}{2}I\omega^2$$

$$I = \frac{1}{3}ML^2 \quad \text{(see Table 8.1)}$$

$$\omega = \sqrt{\frac{3g}{L}}$$

(b)   The tangential and angular velocity are related by Equation (8.7)

$$v_T = \frac{L}{2}\omega \quad \text{(for center of mass)}$$

$$v_T = \frac{1}{2}\sqrt{3gL}$$

* * * * * * * * * * * * * *

8.6 Rolling Without Slipping

The general problem in which an object is turning about an axis, and at the same time, the axis is moving, is very difficult. The only case that can be dealt with in an introductory Physics course is rolling without slipping: Suppose we are at rest, and see a wheel rolling by; its center of mass is moving at constant velocity. The wheel is rolling without slipping if neighboring points on the wheel touch corresponding points on the ground. In other words, in one revolution of the wheel, the circumference unrolls itself on the ground.

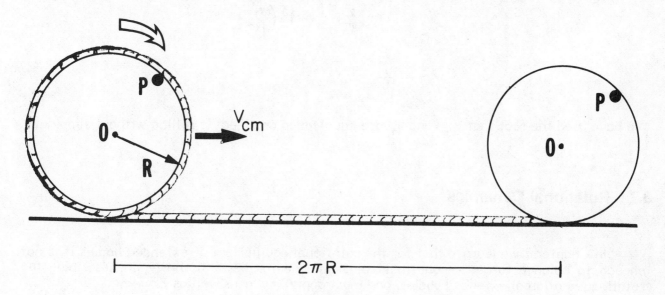

Figure 8.10

From Figure 8.10 we see that the speed of the center of mass O is

$$v_{cm} = \frac{2\pi R}{T}$$

where T is the time for one revolution. In the same time, the point P has moved a distance $2\pi R$ around the point O, so the tangential speed of a point on the rim of the wheel (with respect to the center of mass) is also

$$v_T = \frac{2\pi R}{T}$$

Thus we see that for a wheel that is rolling without slipping, v_T in Eqution (8.7) may be replaced by v_{cm}, and similarly, a_T in Eqution (8.8) may be replaced by a_{cm}. As we shall see shortly, these replacements give us an extra equation which help us solve problems.

When an object is turning about its center of mass with angular velocity ω, and the center of mass is moving with velocity v_{cm} the kinetic energy has two contributions:

$$K = \frac{1}{2}mv_{cm}^2 + \frac{1}{2}I\omega^2 \qquad (8.12)$$

When the object is rolling without slipping, the two terms in Equation (8.12) combine in a neat way. For example suppose our object is a sphere; then we may write

$$K = \frac{1}{2}mv_{cm}^2 + \frac{1}{2}\left(\frac{2}{5}mR^2\right)\omega^2$$

$$= \frac{1}{2}mv_{cm}^2 + \frac{1}{5}mR^2\left(\frac{v_T^2}{R^2}\right)$$

$$= \frac{7}{10}mv_{cm}^2$$

We have used the fact that v_{cm} and v_T are equal when an object is rolling without slipping.

8.7   Rotational Dynamics

In Chapter 7 we learned that for the rotational equilibrium of extended bodies it is not enough to balance forces -- one must also balance torques. Similarly, in describing the rotational motion of extended bodies, one must apply $\Sigma\tau = I\alpha$ as well as $\Sigma F = ma$.

* * * * * * * * * * * * * * *

Example 8-5

A uniform disc of radius R and mass M is mounted on a frictionless horizontal axis as shown in Figure 8.11. A string wrapped around the disc supports a mass m. Find the linear acceleration of the mass m, and the tension in the string.

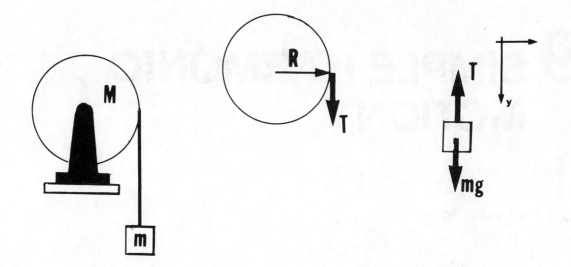

Figure 8.11

Solution: If the string does not slip, the linear acceleration of a point on the rim of the wheel has the same magnitude as the acceleration of the mass m. Apply Newton's Second Law to the mass m:

$$\sum F_y = mg - T = ma$$

The acceleration of mass m is downward, so we take downward as the positive y-direction. Similarly, the wheel has a clockwise angular acceleration, and we take <u>clockwise</u> torques to be positive. Applying $\Sigma\tau = I\alpha$ to the disc yields

$$\sum_{clockwise} \tau = TR = (\tfrac{1}{2}MR^2)\alpha$$

Using $\alpha = \dfrac{a}{R}$ lets us solve the previous equations.

$$mg - T = ma \qquad \text{(force equation)}$$

$$T = \tfrac{1}{2}Ma \qquad \text{(torque equation)}$$

$$mg - \tfrac{1}{2}Ma = ma$$

$$a = \left(\frac{m}{m + \tfrac{1}{2}M}\right)g = \left(\frac{2m}{2m + M}\right)g$$

$$T = \left(\frac{mM}{2m + M}\right)g$$

<u>Note how nicely the R's cancel out of the equation. When this happens, it is usually a sign that you are on the right track!</u>

* * * * * * * * * * * * * * *

9 SIMPLE HARMONIC MOTION

9.1 Introduction

The material in this chapter may <u>look</u> abstract, but really, it is one of the most practical subjects in your entire Physics course. It is an introduction to the theory of vibrations. Vibrations are important in many areas of science and engineering. Molecules vibrate (biology and chemistry), electrical circuits oscillate (electrical engineering), buildings and bridges vibrate (civil engineering), -- even the earth vibrates (geology).

Figure 9.1 The Tacoma Narrows Bridge at Puget Sound. Just four months after its opening on July 1, 1940, a mild gale blowing across the bridge, set it oscillating with such a large amplitude, that the structure gradually weakened, and collapsed within a few hours. The only fatality was a cocker spaniel left in a car by its panicked owner.

"It was only after the Brooklyn and Manhattan Bridges were built that engineers were able to depart from the old, arbitrary methods of figuring lateral stresses, among which those set by wind pressure are especially important. The designer of the Tacoma bridge, Mr. Leon S. Moisseiff, one of the outstanding engineers of our time introduced more scientific methods."

New York Times editorial,
Nov. 9, 1940

In Figure 9.1 we show photographs of the Tacoma bridge designed by Mr. Moisseiff. His understanding of vibratory motion must have been somewhat "shaky".

9.2 Sinusoidal or Simple Harmonic Motion

In vibratory motion an object moves back and forth about an equilibrium point. You are already aware of two mathematical functions which describe such motion -- sine and cosine. <u>Simple</u> <u>harmonic</u> <u>motion</u> or <u>SHM</u> is defined by the equations

$$x = A\cos(\omega t + \delta) \tag{9.1}$$

$$v_x = \frac{dx}{dt} = -\omega A\sin(\omega t + \delta) \tag{9.2}$$

$$a_x = \frac{dv_x}{dt} = -\omega^2\left[A\cos(\omega t + \delta)\right]$$

$$= -\omega^2 x \tag{9.3}$$

where

$$A \equiv amplitude$$

$$\omega \equiv angular\ frequency$$

$$\delta \equiv phase\ angle$$

The frequency and period of the motion are expressed in terms of ω.

$$\omega = 2\pi f = \frac{2\pi}{T}$$

$$T = \frac{1}{f} \tag{9.4}$$

The phase angle δ rarely enters SHM problems, particularly in a non-calculus Physics course. So we defer discussion of δ until Section 9.4. For simplicity, we take δ = 0 in the next few examples.

Memorize Equations (9.1), (9.2), and (9.3) if you must, but here is a case where a little calculus goes a long way. There is really no need to memorize the three equations; v is the first derivative of x with respect to t, and a is the second derivative of x with respect to t.

Whenever you see any one of the Equations (9.1) - (9.3), you must realize that something is performing SHM. For example, the equation

$$a_x = -(glunk)^2 x \tag{9.5}$$

with $(glunk)^2$ a positive constant, describes SHM, where ω = (glunk). In Chapter 5 we considered a mass on the end of a spring which satisfied Hooke's Law:

$$F_x = -kx \tag{9.6}$$

Using Newton's Second Law we get

$$ma_x = -kx$$

$$\tag{9.7}$$

$$a_x = -\frac{k}{m} x$$

So the mass is performing SHM with $\omega = \sqrt{k/m}$.

* * * * * * * * * * * * * * *

Example 9-1

An object is oscillating with SHM described by the equation

$$x = (0.17m)\cos(\omega t)$$

$$\omega = 8\pi \text{ rads/s}$$

(a)   What is t when x = 0.07 m?

(b)   What is x when t = 0.05 s?

(c)   What is the maximum velocity?

(d)   What is the maximum acceleration?

Solution: (a) This is the only part of the problem that causes real trouble. The arc cosine is involved in the solution, and most students are not comfortable with this function.   You will get the wrong answer if you don't stop and think carefully. We don't mean to insult your intelligence when we tell you that we have seen many students make the following grevious error:

$$0.07 = 0.17 \cos (8\pi t)$$

$$t = \frac{0.07}{0.17 \cos 8\pi}$$

Students with math SAT's over 700 generally will not make this error, but they will often ask "How do I get the t out?".

There are two tricks. First realize that the quantity ωt is an angle, and second, that <u>the angle is measured in radians</u>. While you cannot immediately isolate t, you can do the next best thing - you can isolate ωt.

$$x = 0.17 \text{ m } \cos (\omega t)$$

$$\cos (\omega t) = \frac{x}{0.17m}$$

$$\omega t = \cos^{-1}\left(\frac{x}{0.17m}\right)$$

Substituting x = 0.07 m and $\omega = 8\pi$ rad-s^{-1}, we get

$$(8\pi \text{ rad-}s^{-1}) \, t = \cos^{-1}(0.41)$$

$$= 1.15 \text{ rad}$$

$$t = 0.046 \text{ s}$$

We have really determined the first value of t after t = 0 at which x = 0.07 m. Because the motion repeats itself, there are many other values of t at which this happens. These values are obtained by adding $\pm 2\pi$, $\pm 4\pi$, $\pm 6\pi$, etc to the angle ωt, and then resolving for t.

(b)   Here you must remember that the angle is measured in radians.

$$x = 0.17 \text{ m cos} \left[(8\pi \text{ rad-s}^{-1}) (0.05 \text{ s})\right]$$

$$= 0.17 \text{ m cos} (0.4 \pi \text{ rad})$$

$$= 0.053 \text{ m}$$

(c) and (d):   The sine and cosine never have magnitude greater than unity, so Equations (9.2) and (9.3) tell us that the magnitude of the velocity never exceeds ωA, and the magnitude of the acceleration never exceeds $(\omega^2 A)$.

$$v_{max} = \omega A$$

$$= (8 \pi \text{ rad-s}^{-1}) (0.17 \text{ m}) = 4.27 \text{ m/s}$$

$$a_{max} = \omega^2 A$$

$$= (8 \pi \text{ rad-s}^{-1})^2 (0.17 \text{ m}) = 107 \text{ m/s}$$

Note that we have omitted "rad" from the final units of velocity and acceleration. This is because radians have no units. To measure an angle in radians, we divide an arc length by a radius [Eq. (8.6)] — units cancel when you divide a length by a length.

You could write

$$v_{max} = 4.27 \text{ rad-m/s}$$

but it's not the custom, and people would think that you were a bit crazy.

9.3   Combination Problems

The previous material has been all mathematics. In this section we show you how to apply the mathematics of SHM to force and energy problems.

* * * * * * * * * * * * * *

Example 9-2

A block sits on a platform which is moving vertically with simple harmonic motion.

(a)   If the amplitude remains constant at 0.05 m, what is the maximum frequency for which the block and platform remain in contact continuously?

(b) If the frequency remains constant at 2 Hz, at what amplitude of motion will the block and platform separate?

(c) At what point in the motion is the normal force equal to the weight?

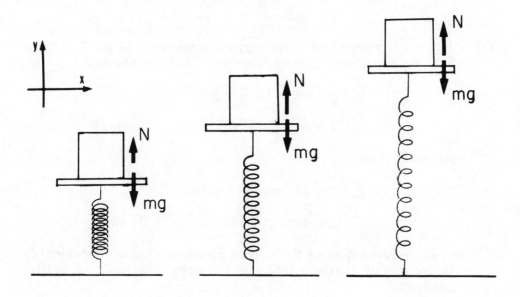

Figure 9.2

Solution: Figures 9.2(a), 9.2(b), and 9.2(c) show the system at minimum extension, intermediate extension, and maximum extension. If the block is to "just" leave the platform, it will do so at the top when the platform jerks down with maximum acceleration. Common sense tells you that this will not happen at the minimum extension, where the platform is pressing up against the block. Applying Newton's Second Law to Figure 9.2(c) yields

$$\sum F_y = mg - N = ma_{max}$$

$$= m\omega^2 A$$

When an object sits on a surface, the surface exerts a normal force on it. When the object leaves the surface, the normal force stops acting. So we use the condition $N = 0$ to solve for the frequency $f = \omega/2\pi$.

$$mg + 0 = m(2\pi f)^2 A$$

$$f = \frac{1}{2\pi}\sqrt{\frac{g}{A}} = \frac{1}{2\pi}\sqrt{\frac{9.8 \text{ m-s}^{-2}}{0.05 \text{ m}}}$$

$$f = 2.23 \text{ Hz}$$

(b)   We proceed as in part (a), except that we solve for the amplitude A rather than the frequency f:

$$A = \frac{g}{(2\pi f)^2} = \frac{9.8 \text{ m-s}^{-2}}{(4\pi \text{ s}^{-1})^2}$$

$$A = 0.248 \text{ m}$$

(c)   This part is very instructive; as we have seen, at the "top", the SHM acceleration has magnitude $\omega^2 A$ and we may write

$$\sum F_y = mg - N = m\omega^2 A$$

$$N = mg - m\omega^2 A \qquad (N < mg)$$

At the bottom

$$\sum F_y = N - mg = m\omega^2 A$$

$$N = mg + m\omega^2 A \qquad (N > mg)$$

N will equal mg when the simple harmonic motion acceleration is zero; this happens halfway between the top and bottom positions.

* * * * * * * * * * * * * *

Example 9-3

A block is on a horizontal surface which is moving horizontally with SHM of frequency 3.5 Hz. The coefficient of static friction between block and plane is 0.30. What is the maximum amplitude such that the block does not slip along the surface?

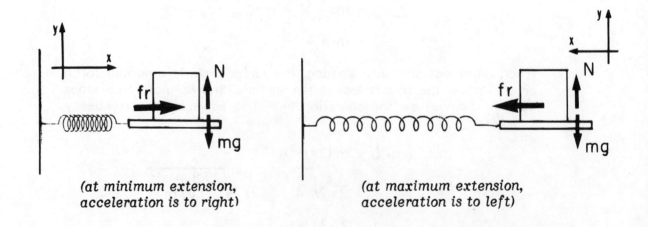

(at minimum extension,
acceleration is to right)

(at maximum extension,
acceleration is to left)

Figure 9-3

Solution: The slipping will occur at either maximum or minimum extension when the acceleration is greatest, and the surface tends to pull out from under the block. The force of static friction is the only force in the horizontal direction, and it keeps the block on the plane. As long as the block is in contact with the surface, we may write (for both maximum and minimum extension)

$$\sum F_x = fr = m \omega^2 A$$

$$\sum F_y = N - mg = 0$$

$$A = \frac{fr}{m\omega^2}$$

As we jack up the amplitude, the force of static friction gets larger and larger to keep up with the ma term. But there is a maximum value that it can attain, and for larger values of A slipping will occur. The maximum value of a is obtained by setting

$$fr = \mu_s N$$

$$= \mu_s mg$$

$$A = \frac{\mu_s g}{(2\pi f)^2} = \frac{(0.3)(9.8 \text{ m-s}^{-2})}{(7\pi \text{s}^{-1})^2}$$

$$A = 6.1 \times 10^{-3} \text{m}$$

* * * * * * * * * * * * * *

Example 9-4

A particle is performing SHM with amplitude A. What is the magnitude of the displacement when the kinetic energy K is twice the potential energy U?

Solution: The total mechanical energy E of the oscillating particle is

$$E = \frac{1}{2} mv^2 + \frac{1}{2} kx^2$$

$$K = 2U$$

$$\frac{1}{2} mv^2 = 2\left(\frac{1}{2} kx^2\right)$$

$$E = \frac{3}{2} kx^2$$

Now we use the fact that the total mechanical energy is equal to the maximum potential energy ($\frac{1}{2}kA^2$):

$$\frac{1}{2}kA^2 = \frac{3}{2}kx^2$$

$$x = \frac{1}{\sqrt{3}}A$$

There are many variations of this type of problem. Usually the object is to eliminate the kinetic energy K in terms of either E or U. For example, if you are told that $U = 1/4\, E$, you should quickly realize that $K = 3/4\, E$.

10 WAVE MOTION

10.1   Introduction

A wave is a means of transferring energy over large distances without appreciable transfer of mass. Waves that we usually deal with are classified as either mechanical or electromagnetic. Mechanical waves involve vibrations in a material medium and include sound, ocean waves, and seismic waves. On the other hand, radar and radio waves, light, X-rays, and gamma rays, are examples of electromagnetic waves. Such waves do not require a medium for their propagation, and can be considered concentrations of pure electrical and magnetic energy spreading out through space with the speed of light. In this chapter we do not discuss the nature of specific kinds of waves; rather we concentrate on more general properties that all waves have in common.

An understanding of the nature of waves and of wave phenomena is important in many areas of science and engineering. For example, whole industries are based on the branches of Physics known as Optics and Acoustics which deal with the properties of light and sound waves respectively. Also, such waves are of fundamental importance in sciences such as Biology and Psychology, since most animals (including humans) communicate and obtain information about their environment by detecting light and/or sound.

In recent years human beings have learned to use many types of electromagnetic waves other than light. The applications to radar, radio and television are well known. The use of X-rays in medical diagnostics and cancer therapy are also familiar. Less known is the fact that measurements of very small distances are done almost exclusively using the phenomena of diffraction and interference. X-ray diffraction (to be discussed in Chapter 18) is an invaluable tool of the biochemist in determining the distances between atoms, and thus the structure of complex organic molecules such as DNA, insulin, etc.

For most students the concepts of wave motion are not intuitive -- past experience doesn't often help. The problems are not very difficult if you understand the concepts. So in this chapter we concentrate more on concepts, and less on problems.

10.2   Traveling Waves

The most obvious way of transferring energy from one point to another is by direct transfer of matter. For example, if we throw an object of mass m and it arrives at its destination with velocity v, it will deposit energy ½ mv at that point. Transfer of energy by a wave occurs in a much different manner, as shall now be explained for the case of a continuous medium such as a string, air, or water. If we imagine such substances to be divided into many small portions or elements, each such element is coupled to its neighbor by forces which tend to resist deformation. Thus if a portion of a string (or a layer of air or water) is set into motion, the motion will be transmitted to the adjoining portions, and so forth. The net result is that a pattern carrying energy which we call a traveling wave, moves out into the medium from the original source of the motion. There are two key features of this mode of energy transfer:

1.   The moving pattern which carries the energy more or less maintains its shape, and moves with a characteristic velocity called the propagation or wave velocity.

2.   While there is an organized transfer of energy over large distances, there is negligible transfer of matter over such distances.

Students tend to confuse the wave velocity with the velocities of the elements of the medium as the wave passes by. In Example 10-1 we show that the wave and particle velocities can have very different behaviors.

* * * * * * * * * * * * * *

Example 10-1

A dramatic case of energy transfer via waves occurs in the case of tsunami, often called tidal waves in popular usage. The latter terminology is misleading since these waves have nothing to do with tidal phenomena; it is more like a magnified version of what happens when you drop a stone in a lake. A tsunami is a seismic sea wave set up when a submarine earthquake abruptly raises or lowers portions of the sea bottom. The surface of the water takes a certain shape or pattern depending on the nature of the disturbance, and then the shape/pattern moves horizontally across the ocean at typical speeds of 800 km/h. Further statistics which may surprise you are that a characteristic length of such disturbances is 150 km, and that in deep water (for example, in mid-Pacific Ocean) a typical height is only about 0.5 m! Ships in mid-ocean will hardly notice these waves.

Disastrous events begin to occur when the waves approach shore, and all the energy that moved thousands of cubic meters of water in the open ocean is concentrated on moving relatively few cubic meters of water up a shallow shore. The water may build up to heights of over 30 m flooding coastal areas with waves of great force. Tsunami have caused great damage and much loss of life throughout human history. In 1883 the tsunami, which were created when the famous volcano at Krakatoa erupted, drowned more than 30,000 people on the islands of Java and Sumatra.

* * * * * * * * * * * * * * *

Figure 10.1 describes a tsunami with characteristic length and height. Suppose that it is moving to the right with a speed of 800 km/h, and at t = 0 it strikes a boat located at x = 200 km (see Figure 10.1).

(a) Does the boat have reasonable dimensions relative to the wave shape?

(b) Calculate the average velocity with which the boat is lifted from sea level to the crest of the tsunami.

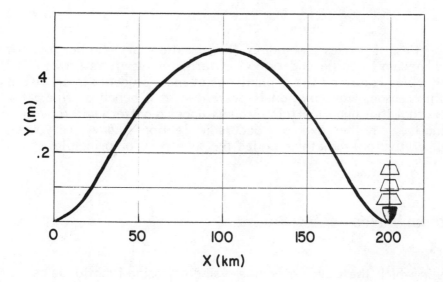

Figure 10.1

Solution: (a) Please note that the scale is different on the two axes. You can't go wrong on this part of the problem. If you feel that a boat that is approximately 0.17 m high and has sails 20 km wide is reasonable, answer <u>yes</u>; if not, answer <u>no</u>!

(b) We are aware of your urge to plug numbers in some equation, but for the moment, exhibit self control and <u>think</u> about what happens as the wave moves toward the boat from the left.

The crest of the wave starts 100 km from the boat and moves toward it at a speed of 800 km/h. Therefore it will arrive in 1/8 h.

The surface of the water assumes the shape of the wave pattern. Since the boat floats on top of the water, it rises as the wave approaches the boat.

When the crest has arrived, the boat will be at its maximum height or 0.5 m above sea level. It will have risen 0.5 m in 1/8 h, so

$$v_{av} = \frac{0.5\ m}{1/8\ h} = 4.0\ \frac{m}{h}$$

Clearly the wave described above could not upend an ocean liner in mid-ocean. (Did you see the movie entitled "The Poseidon Adventure"?)

Note that we have our answer without using a single equation from your textbook chapter on wave motion! Dependence on equations is a particularly bad habit in the case of wave motion problems.

$$* * * * * * * * * * * * * *$$

The properties of waves (including the question of whether or not a wave will propagate in a medium) can be understood in terms of Newton's Laws. Unfortunately the mathematics involved is beyond the capabilities of most students in an introductory Physics course. For this reason, wave motion is presented as a bunch of rules; for example, you must accept the fact that under certain circumstances a pattern will move in a medium with a definite velocity. Another example occurs in Section 10.4 where you will be asked to accept (without motivaton) something called the superposition principle.

10.3   Sinusoidal Waves

The case in which the wave pattern is a sine or cosine function is especially important in Physics. For example, in the case of light, a sine wave of a single frequency describes a single color; in the case of sound, such a wave describes a single note. A sinusoidal wave in one dimension is described by the mathematical function

$$y = A\cos(\omega t - kx + \phi) \tag{10.1}$$

where t is time, x is position along the x-axis, and A, ω, ϕ, and k are constants which you shall learn about shortly. The minus sign corresponds to a wave moving to the right; the plus sign indicates a wave moving to the left. In Equation (10.1) we have tried to adopt the most common notation. If the author of your text likes to use sine rather than cosine, or $- \phi$ rather than $+ \phi$, don't worry. These differences will not affect the explanations that follow.

There is a reason that the mathematics of wave motion causes trouble for beginning students. When you use Equation (10.1), it is probably the first time in your life that you have really worked with a complicated function of two variables. It would require three axes to study the function graphically (a y-axis, a x-axis, and a t-axis). Most of us have difficulty with curves in three dimensions. So the usual approach is to consider two different "pictures" of the wave:

I.   The y-t picture (keep x constant)

In this case we focus on a particular value of x (a fixed point in space) and watch y as a function of time t. If we were dealing with transverse waves on a string we would be watching a small piece of string at a particular value of x, going up and down. Suppose x is constant in Equation (10.1); then the equation has the form

$$y = A\cos(\omega t + \delta)$$
$$\delta = \phi - kx$$

(10.2)

where δ is a constant because x is a constant. So we have proven that every piece of the string moves up and down with simple harmonic motion [see Eq. (9.1)]. The quantity ω has the same significance that it did in Chapter 9; it is the angular frequency.

2.   The y-x picture (keep t constant)

Keeping t constant means that you are taking a snapshot of the wave, looking at the values of y at all points x at one time. If we now plot y as a function of x (with t constant) we have introduced something new for you. For example suppose we take ϕ to be zero, and fix t at the value t = 0; Equation (10.1) becomes

$$y = A\cos kx$$
$$k \equiv \frac{2\pi}{\lambda} \quad (definition \; of \; wavelength \; \lambda)$$

(10.3)

The wavelength λ is the distance in which the cosine curve repeats itself.

The y-t and y-x pictures are not independent. The angular frequency ω and the wave number k are related to the wave velocity v by

$$\omega = kv$$

(10.4a)

and therefore the relation between the constants f, T, λ, and v is

$$v = f\lambda = \frac{\lambda}{T}$$

(10.4b)

Equations (10.4 a) and (10.4 b) play such an important role in your studies of wave motion, that we shall refer to them as the <u>fundamental equations of wave motion</u>. MEMORIZE THESE EQUATIONS!

We apply Equations (10.1) - (10.4) in Examples 10-2 and 10-3 below. Example 10-2 is the simpler of the two problems, and the most likely type to appear on an examination. It is straightforward, except for one stumbling block: don't confuse the wave velocity and the "maximum transverse speed".

* * * * * * * * * * * * * * *

Example 10-2

(a)   Write an expression describing a transverse wave traveling on a string in the +y direction with a velocity of 1.25 m/s, a period of 0.0025 s, and having an amplitude of 0.5 cm. Take the transverse direction to be the z direction, and use a cosine function with zero phase angle.

(b)   What is the maximum transverse speed of a point on the string?

Solution: (a) Calculate the amplitude, angular frequency, and wave number. Substitute these values into Equation (10.1), using the minus sign because the wave is traveling in the "plus" y direction.

$$A = 0.5 \text{ cm} = 0.005 \text{ m}$$

$$\omega = \frac{2\pi}{T} = \frac{2\pi}{0.0025 \text{ s}} = 2513 \text{ s}^{-1}$$

$$k = \frac{\omega}{v} = \frac{2513 \text{ s}^{-1}}{1.25 \text{ m/s}} = 2010 \text{ m}^{-1}$$

$$z = A \cos (2513 \, t - 2010 \, y)$$

(b)   The maximum transverse speed of a point on the string is obtained by using Equation (10.2), which says that every point on the string is performing simple harmonic motion with angular frequency ω. From our studies of simple harmonic motion, we know that the maximum speed in such motion is

$$v_{z,max} \ = \ \omega A \ = \ (2513 \ s^{-1})(0.5 \ m)$$

$$= \ 1257 \ m/s$$

The answer is not the wave velocity of 1.25 m/s !

* * * * * * * * * * * * * * *

Example 10-3

The pattern shown in Figure 10.2 represents the pressure variation at $t = 0$, in a sound wave which is moving to the right with a velocity of 330 m/s.

(a) At which points in space is the pressure rising as a function of time?

(b) Plot the pressure variation at $x = 20$ m as a function of time.

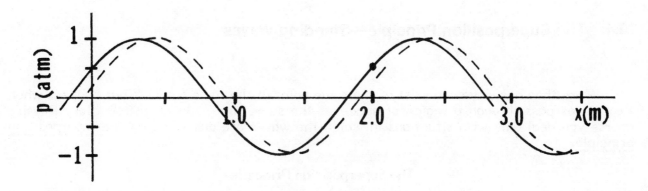

Figure 10.2 Wave at $t = 0$.

Solution: (a) Remember that the curve in Figure 10.2 gives the pressure at each point x at $t = 0$. At a slightly later time the curve has shifted to the right as indicated by the dashed curve. Where the dashed curve is below the solid curve the pressure is falling, and where the dashed curve is above the solid curve, the pressure is rising. This is the easy way to do part (a). Thus the answer is: from $x = 0.5$ m to 1.5 m, and from 2.5 m to 3.5 m, etc.

(b) The easy way to do part (b) is to set up a y-t graph as shown in Figure 10.3

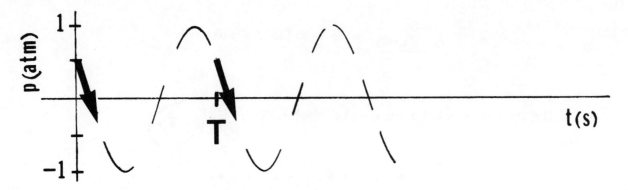

Figure 10.3 Wave at x = 2.0 m.

We know two pieces of information from Figure 10.2 and part (a). The pressure at x = 2.0 m and t = 0 is 0.05 atm, and furthermore the pressure is falling. For t greater than zero, the values of y will trace out a "sine" curve that repeats itself in a period T.

$$\lambda = 2.0 \text{ m (from Figure 10.2)}$$

$$T = \frac{\lambda}{v} = \frac{2.0 \text{ m}}{330 \text{ m/s}}$$

$$T = 0.0061 \text{ s}$$

* * * * * * * * * * * * * * *

10.4   The Superposition Principle—Standing Waves

More than one wave can exist at the same point at the same time. What happens when two waves pass through a region of space at the same time? In an introductory Physics course you deal only with situations in which the waves are described by the <u>superposition principle</u>.

The Superposition Principle

If at any instant two or move waves simultaneously exist at a point, the displacement of the point is the sum of the displacements the point would have with each wave separately.

<u>Interference</u> refers to the physical effects of superimposing two or more wave trains. For example, two sine waves of the same frequency, speed, and amplitude, traveling in opposite directions on a string result in <u>standing waves</u>.

$$y = A sin(kx - \omega t) + A sin(kx + \omega t) \tag{10.5a}$$

Using the trigonometric identities

$$\sin (D \pm E) = \sin D \cos E \pm \cos D \sin E$$

we get

$$y = A \sin kx \cos \omega t - A \cos kx \sin \omega t$$

$$+ A \sin kx \cos \omega t + A \cos kx \sin \omega t$$

$$y = 2A \sin kx \cos \omega t = 2A \sin\left(\frac{2\pi}{\lambda}x\right)\cos\omega t \qquad (10.5b)$$

Equation (10.5) tells us that there are points on the string which are at rest for all times t. These points are called nodes; they occur when

$$x = 0, \frac{\lambda}{2}, \lambda, \frac{3}{2}\lambda, 2\lambda \ldots \qquad (10.6)$$

Standing waves may be set up on a string fixed at both ends by plucking it. In Figure 10.4 we illustrate the four longest wavelengths that will propagate on a string (actually a flexible rubber tube) of length L. Rather than plucking the "string", the experimenter in Figure 10.4 gets a standing wave by wiggling the rubber tube at the correct frequency.

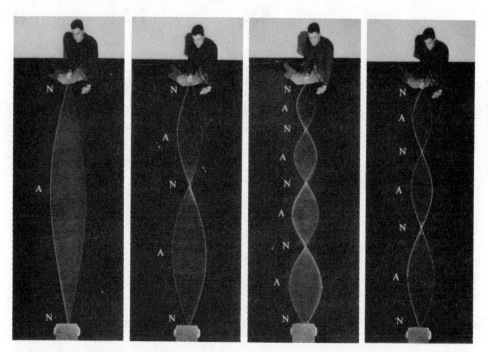

Figure 10.4 Standing waves. A long flexible rubber tube is rigidly attached at one end. If the experimenter wiggles the other end at the correct frequency, a standing wave is produced. Each particular pattern is associated with a definite frequency. In this situation the tension remains constant. N and A show nodes and antinodes, respectively. From Physical Science Study Committee. <u>Physics</u>, D.C. Heath, Lexington, MA, 1960, with permission.

Remember Figure 10.4. Having this picture in your mind will often prove more useful on examinations than being able to work with Equation (10.5).

* * * * * * * * * * * * * *

Example 10-4

A string is stretched between two walls and tied at points A and B as shown below. Paper "riders" are placed on the string at points C and D, and standing waves are set up. (See Figure 10.5.)

What are the three longest wavelengths for which the "riders" will remain at rest?

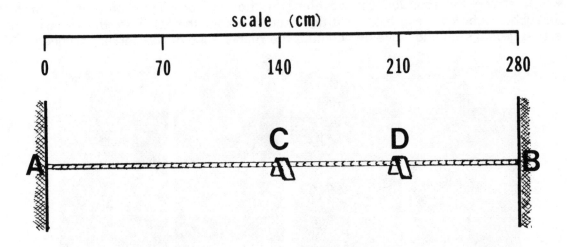

Figure 10.5

Solution: The paper riders will remain at rest if they are sitting on nodes. So we are interested in standing waves which produce nodes at points C and D. The two longest wavelengths are shown in Figures 10.6(a) and 10.6(b). The darkened line <u>represents</u> the wavelength of one of the waves traveling on the string. At any instant we <u>see</u> a superposition of the waves on the string and cannot distinguish the traveling waves that make up the standing wave. Still, the diagram in Figure 10.6 serves as a mnemonic to get the wavelength (i.e., the darkened line). In Figure (10.6a) the "wavelength" occupies <u>two</u> of <u>four</u> loops, so λ = ½L, and in Figure (10.6b) the wavelength occupies <u>two</u> of <u>eight</u> loops, so λ = ¼L.

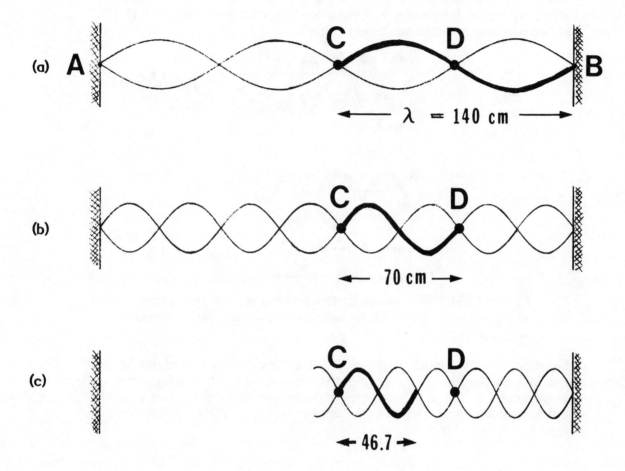

Figure 10.6 Schematic representation of standing waves on string.

It's the third longest wavelength that fools the majority of students. They're already in the rhythm of halving wavelengths, so they get an answer of 35 cm, which is wrong! The third longest wavelength has three loops between points C and D, so

$$\lambda = \frac{2}{3}(70 \text{ cm})$$

$$= 46.7 \text{ cm}$$

* * * * * * * * * * * * * * *

Example 10-5

A stretched string of length L, fixed at both ends, is observed to vibrate in five equal segments when driven by a 550 Hz oscillator. What oscillator frequency will set up a standing wave such that the string vibrates in three equal segments?

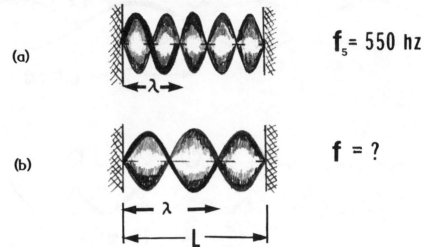

(a) $f_5 = 550$ hz

(b) $f = ?$

Figure 10.7(a) The wavelength occupies two of five loops, so $\lambda_5 = 2L/5$. **(b)** The wavelength occupies two of three loops, so $\lambda_3 = 2L/3$.

Solution: From Figure 10.7 we see that the five segment vibration has a wavelength of 2L/5, and the three segment vibraton has a wavelength of 2L/3. The velocity of the wave is the same in both cases, so we may obtain our answer by using the fundamental equation of wave motion.

$$f_5 = \frac{1}{\lambda_5} v$$

(A) $$550 = \frac{v}{2L/5} = \frac{5}{2}\frac{v}{L}$$

(B) $$f_3 = \frac{v}{2L/3} = \frac{3}{2}\frac{v}{L}$$

Divide (A) by (B): $$f_3 = \frac{3}{5}(550)\ \text{Hz} = 330\ \text{Hz}$$

Note that we obtained our result using few equations. Knowing the contents of Figure 10.4 was the important thing! It also helps to know what to expect. The <u>lower</u> frequency has the <u>longer</u> wavelength, so f_3 should be less than f_5. If, for example, you had divided incorrectly and obtained

$$f_3 = \frac{5}{3}(550)\ \text{Hz} = 917\ \text{Hz},$$

you should immediately know that you made an error!

* * * * * * * * * * * * * *

You will encounter other types of standing waves, but simple mnemonics of the type we have considered, will also work in these cases. Some examples are

(a)   Sound waves in an organ pipe open at both ends.

(b)   Sound waves in an organ pipe open at one end.

(c)   Waves on a string fixed at one end, and being whipped with maximum displacement at the other end.

Case (a) is treated in Figure 10.8(a); the standing wavelengths are the same as those on a string fixed at both ends.

Case (c) has the same standing wavelengths as case (b) which is illustrated in Figure 10.8(b).

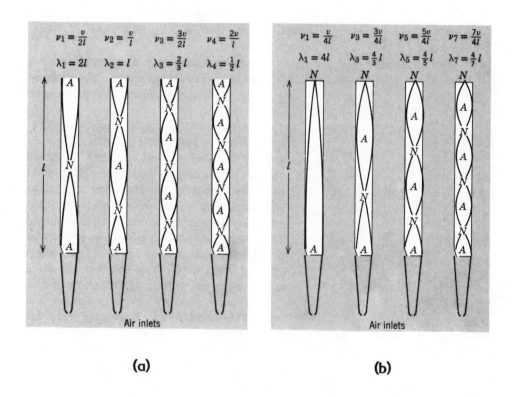

(a)   (b)

Figure 10.8 (a) The first four modes of an open organ pipe. The distance from the center line of the pipe to the right lines drawn inside the pipe represent the displacement amplitude at each place. N and A mark the locations of the displacement nodes and antinodes. Note that both ends of the pipe are open. **(b)** The first four modes of vibration of a closed organ pipe. Notice that the evennumbered harmonics are absent and the upper end

10.5   Interference Between Two Coherent Sound Sources

In Figure 10.9 we show two speakers emitting sinusoidal sound waves of one frequency in phase. By this statement we mean that the two waves are described by the same sine or cosine functions as the two speakers (there are no phase differences). In general, if waves emitted from two sources have a definite phase relation (there does not have to be zero phase differences), the sources are said to be coherent.

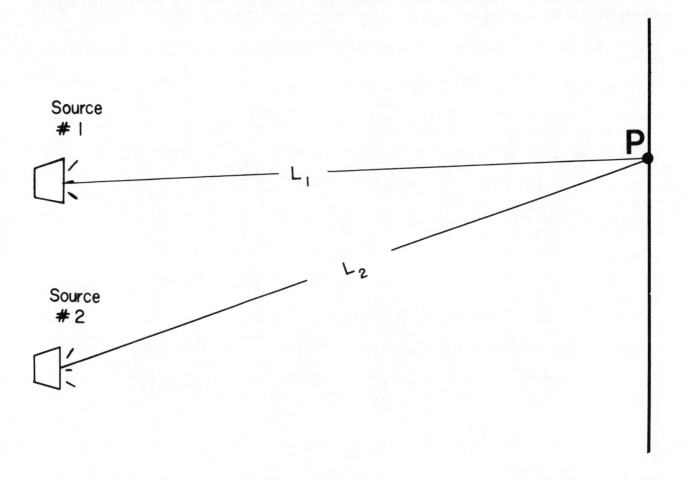

Figure 10.9

Amazing things can happen at point P when the waves from the two sources superimpose. For example, if the amplitudes are equal it is possible to hear nothing! We explain this phenomenon: Crudely speaking, coherence means that waves leaving sources #1 and #2 are always doing the same thing as a function of time. If the paths L_1 and L_2 are equal, they will be doing the same thing at the same time at point P, and add up to give a resultant sinusoidal wave with maximum possible amplitude as shown in Figure 10.10.

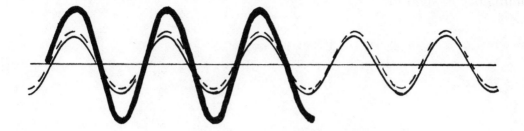

Figure 10.10 Maximum constructive interference.

Figure 10.10 also describes the situation if L_1 and L_2 differ by an <u>integral number of</u> <u>wavelengths</u>. (λ, 2λ, 3λ, etc.) since a traveling sinusoidal wave repeats its behavior at points separated by one wavelength.

On the other hand, suppose that L_1 and L_2 differ by $\lambda/2$, $3\lambda/2$, $5\lambda/2$, etc. Now, the waves come into point P doing the "opposite thing" as a function of time, and Figure 10.11 describes the situation.

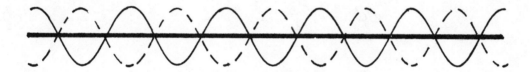

Figure 10.11 Total destructive interference.

The resultant wave has <u>zero amplitude</u> for all time. The intensity of the sound at point P is the energy passing through a unit area in a unit time, and is proportional to the square of the amplitude. <u>Zero amplitude</u> means you hear nothing at point P!

The previous discussion may be summarized in the following useful manner.

Intensity maxima occur when the path difference

$$\Delta x = 0, \lambda, 2\lambda \ldots \tag{10.7}$$

or the phase difference

$$\Delta \phi = 0, 2\pi, 4\pi \ldots \tag{10.8}$$

Intensity minima occur when the

$$\Delta x = \frac{\lambda}{2}, \frac{3\lambda}{2}, \frac{5\lambda}{2} \dots \tag{10.9}$$

or the phase difference

$$\Delta\phi = \pi, 3\pi, 5\pi \dots \tag{10.10}$$

The general relation between path difference and phase difference is

$$\Delta\phi = \frac{2\pi}{\lambda}\Delta x \tag{10.11}$$

The most extensive application of these equations will occur in Chapter 17 on Physical Optics, but occasional questions of this type appear on mechanics exams.

* * * * * * * * * * * * * *

Example 10-6

In Figure 10.9, both sources are emitting sound waves of frequency f = 2500 Hz. Suppose that L_1 = 20 m and L_2 = 20.5 m. Does one hear maximum or minimum intensity at point P?

Solution: First calculate the path difference

$$\Delta x = (L_2 - L_1)$$

$$= .5 \, m$$

The wavelength of the sound is obtained from the fundamental equation of wave motion. Use 330 m/s for the velocity of the sound.

$$v = f\lambda$$

$$330 \text{ m-s}^{-1} = 2500 \text{ s}^{-1}\lambda$$

$$\lambda = 0.13 \text{ m}$$

$$\frac{\Delta x}{\lambda} = \frac{.5 \text{ m}}{0.12 \text{ m}} = 4.17$$

$$\Delta x = 4.17 \lambda$$

So we have neither an intensity minimum, nor an intensity maximum!

* * * * * * * * * * * * * * *

11 HEAT AND THERMODYNAMICS

Heat is work and work is heat.

*Heat cannot of itself pass
from one body to a hotter body.*

Heat won't pass from a cooler to a hotter.

*You can try it if you like,
but you'd far better notter.*

*Cause the cold in the cooler
will get hotter as a ruler.*

*Cause the hotter body's heat
will pass to the cooler.*

Flanders and Swann
Angel Records 1961

11.1 Introduction—Internal Energy

The concept of internal (or thermal) energy was developed at the end of the 19th century. It is the energy of motion and the potential energy of the zillions of atoms in a material substance, such as a piece of iron, or a pan of water, or the air. It must be pointed out immediately that this energy is <u>very different</u> from the kinetic energy of an iron hammerhead swinging through the air, or a gush of water coming from a fire hose, or a gust of wind. The difference is that in the latter examples, the kinetic energy is organized energy. For example, when the hammerhead is being swung, all the atoms in the iron are being forced to move together, in the same direction. Because of this organized motion, the hammerhead can do a great deal of work on hitting a nail. On the other hand, a hot piece of metal can do very little work when held to a nail, even though the sum of the kinetic energies of all the iron atoms in the hot hammerhead at rest may be greater than that of a sharply swung hammerhead. This is because the kinetic energies which make up the thermal energy of a body are disorganized. It's like throwing a bag of popcorn (organized kinetic energy) versus letting them pop around inside the popper (disorganized kinetic energy). Popcorn popping very quickly has lots of thermal energy, but can do little work.

The internal energy of a substance is represented by the symbol U. In general it is a complicated function of the temperature. Fortunately, the "substance" one mostly deals with in an introductory Physics course is an <u>ideal gas</u>. For the case of an ideal monatomic gas

$$U = \frac{3}{2}NkT$$

$$N \equiv number\ of\ molecules \tag{11.1}$$

$$k \equiv Boltzman's\ constant$$

$$T \equiv Absolute\ temperature$$

or

$$U = \frac{3}{2}nRT \tag{11.2}$$

$$n \equiv number\ of\ moles^{*}$$

$$R \equiv universal\ gas\ constant$$

Equation (11.1) gives some feeling for the physical quantity temperature. If you solve this equation for the temperature, you get

$$T = \frac{2}{3k}\left(\frac{U}{N}\right) \tag{11.3}$$

So the absolute temperature is proportional to U/N which is the average internal energy per molecule.

---

* A *mole* (abbr. mol) of any substance is the amount of the substance that contains a specified number of elementary entities, namely, 6.022045×10^{23}, called the Avogadro constant. This number is the result of the defining relation that one mole of carbon atoms (actually, of the isotope ^{12}C) shall have a mass of 12 g, exactly. The *gram molecular weight* M of a substance is the number of grams per mole of that substance. Thus the gram molecular weight of ordinary oxygen molecules is 32.0 g/mol. Although the mole is an amount of substance, we cannot translate it into mass, as grams, until we specify what the elementary entity is; it may be atoms, molecules, ions, electrons, other particles, or specified groups of such particles.

Example 11.1

Calculate the internal energy of one mole of helium gas at the freezing point of water. R = 8.31 J/mol-K.

Solution: Helium behaves like a monatomic ideal gas so we may use Eq. (11.2) to get U. The freezing point of ice is 273 K, so

$$U = \frac{3}{2} (1 \text{ mol}) (8.31 \frac{J}{mol\,K}) (273 \text{ K})$$

$$= 3403 \text{ J}$$

We hope that you didn't use T = 0° C and get

$$U = 0 !$$

A major error in problems involving the ideal gas law (PV = nRT) and derived results such as Eq. (11.2) is to use degrees celsius rather than degrees kelvin. Remember

$$T \text{ (kelvin)} = T \text{ (celcius)} + 273$$

* * * * * * * * * * * * * *

Internal energy and temperature are two of the three important new concepts that you must learn in this chapter. The third concept is heat. When a body at a higher temperature is placed in contact with another body at a lower temperature, energy flows from the hotter body to the colder body. This energy is called heat and is represented by the symbol Q.

$$Q \equiv heat$$

The First Law of Thermodynamics ties all these ideas together. It is an expression of the principle of conservation of energy which includes heat and internal energy:

$$Q = \Delta U + W$$

$$W = \sum P\Delta V \quad (work \ by \ the \ system) \qquad (11.4)$$

$$\Delta U \equiv change \ in \ internal \ energy,$$

where P is the pressure, and V is the volume. Equation (11.4) says that if we add heat to a substance, part of the heat will increase the internal energy of the substance. But in general something else will happen also; the substance will expand and do work against the external environment. In the next section, we apply the First Law of Thermodynamics to the four processes which one usually encounters in an introductory Physics course. These are

constant volume processes

constant pressure processes

constant temperature or isothermal processes

adiabatic processes

11.2   Applications of the First Law of Thermodynamics

A.   Constant Volume Processes

The constant volume processes is the simplest of the thermodynamic processes that we shall consider. The nicest thing about constant volume processes is that no work is done.

$$W = \sum P\Delta V$$

$$= 0 \quad (for \ V \ constant)$$

(11.5)

If the volume remains constant, all the ΔV's are zero; and so W is zero. Thus all the heat added at constant volume goes into internal energy.

$$Q = \Delta U \quad (at \ constant \ volume)$$

* * * * * * * * * * * * * *

Example 11-2

The temperature of 2.5 moles of an ideal monatomic gas rises from 150 K° to 350 K°, while the volume remains constant. For an ideal gas the specific heat per mole at constant volume is 3R/2, where R = 8.31 J/mol-K.

(a) How much work is done?

(b) How much heat is added to (or subtracted from) the system?

(c) What is the change in internal energy?

Solution: (a) The answer is zero; we have already explained that no work is done in a constant volume process.

(b) Knowing the specific heat (at constant volume) and the initial and final temperature, we can calculate the heat added to the system. In this case it is a positive number.

$$Q = n \, C_V \, \Delta T$$

$$= (2.5 \text{ mol}) \frac{3}{2} (8.31 \, \frac{J}{\text{mol K}}) (200 \text{ K})$$

$$= 6233 \text{ J}$$

(c) $Q = \Delta U$ in a constant volume process, so the answer is the same as in part (b).

* * * * * * * * * * * * * *

B. Constant Pressure Processes

When heat is added to a substance at constant pressure, the internal energy increases, and at the same time the substance expands and does work. Because the pressure P is constant, it may be factored out of the sum in Equation (11.4) and we get

$$Q = \Delta U + P(V_f - V_i) \qquad (\text{at constant pressure}) \qquad (11.6)$$

where v_f is the final volume and v_i is the initial volume.

* * * * * * * * * * * * * *

Example 11-3

The temperature of 2.5 moles of an ideal monatomic gas rises from 150 K to 350 K while the pressure remains constant at 10,000 N/m².

    (a) Calculate the work done?

    (b) Calculate the change in internal energy?

    (c) What is the heat added to (or subtracted from) the substance?

Solution: (a) We use the ideal gas law to get the final and initial volumes. The work may then be calculated. The ideal gas law states that $PV = nRT$, so

$$V_i = \frac{nRT_i}{P_i} = \frac{(2.5 \text{ mol}) (8.31 \text{ J/mol K}) (150 \text{ K})}{10,000 \text{ N/m}^2} \times \frac{1 \text{ N-m}}{J} = 0.312 \text{ m}^3$$

Similarly,

$$V_f = \frac{nRT_f}{P_f} = \frac{(2.5) (8.31) (350)}{10,000} \text{ m}^3 = 0.727 \text{ m}^3$$

$$W = P (V_f - V_i) = 10,000 (0.727 - 0.312) \text{ J} = 4150 \text{ J}$$

    (b) The important thing to realize is that the change in internal energy is the same as in part (b) of Example 11-1. The change in internal energy depends only on the number of moles and the temperature change.

$$\Delta U = 6,233 \text{ J}$$

    (c) Just add the results of parts (a) and (b).

$$Q = \Delta U + W$$

$$= 6,233 \text{ J} + 4,150 \text{ J}$$

$$= 10,380 \text{ J}$$

* * * * * * * * * * * * * *

C.   Isothermal Processes

Isothermal means constant temperature -- for an ideal gas, constant temperature means no change in internal energy. For this case the First Law gives

$$Q = \Delta U + W$$
$$\searrow 0$$

$$Q = W \quad (in \ an \ isothermal \ process\ ) \tag{11.7}$$

So it is really only for an isothermal process that ♪ heat is work, and work is heat ♫. Don't confuse heat and temperature. In an isothermal process, heat enters or leaves the system, even though the temperature does not change.

For an ideal gas, the pressure, volume, temperature, and number of moles are related by

$$PV = nRT \tag{11.8}$$

If the number of moles n remains constant

$$\frac{P_i V_i}{T_i} = \frac{P_f V_f}{T_f} \tag{11.9}$$

If in addition, the temperature remains constant (isothermal process)

$$P_i V_i = P_f V_f \quad (Boyle's \ Law) \tag{11.10}$$

D.   Adiabatic Processes

In an adiabatic process Q = 0; no heat enters or leaves the system, so the temperature must change. In this case the First Law of Thermodynamics yields

$$Q = \Delta U + W$$
$$0 = \Delta U + W$$
$$\Delta U = -W \quad (in \ an \ adiabatic \ process\ ) \tag{11.11}$$

Do not confuse heat and temperature. No heat transfer does not mean no temperature change. In fact, it means exactly the opposite!

For an ideal gas, the following formulas are valid during an adiabatic process:

$$\frac{P_i V_i}{T_i} = \frac{P_f V_f}{T_f} \quad \text{(always true for an ideal gas if number of moles is constant)}$$

$$P_i V_i^\gamma = P_f V_f^\gamma \quad (\gamma = \frac{C_p}{C_v}, \text{ for a monotomic ideal gas } \gamma = \frac{5}{3}) \tag{11.12}$$

$$\Delta U = -W$$

Don't use Boyle's Law in an adiabatic process!

* * * * * * * * * * * * * *

Example 11-4

In terms of the First Law of Thermodynamics, explain why a bicycle pump heats up when you are pumping air into a bicycle tire.

Solution: The downstroke (when air is being pushed into the tire) happens very quickly, and heat doesn't have much time to enter or leave the system. So the process is approximately adiabatic, and Equation (11.12) holds:

$$\Delta U = -W$$

The important point here is to remember that work is positive when the system does work, but it is negative when work is done on the system. In this case we are doing work on the gas, so W is negative and ΔU is positive in the above equation. A positive ΔU means an increase in temperature. (If we do work on the gas, we increase its internal energy, hence its temperature.)

* * * * * * * * * * * * * *

11.3 Closed Loop Processes

In a real engine the same process happens over and over again. The real cyclic processes that occur in an engine can often be approximated by closed loop diagrams in the

P-V plane, using the simple thermodynamic processes described in Section 11.2. We show such a diagram in Figure 11.1. The closed loop is bounded by a constant volume line, a constant pressure line, and an adiabat (a curve along which no heat enters or leaves the system).

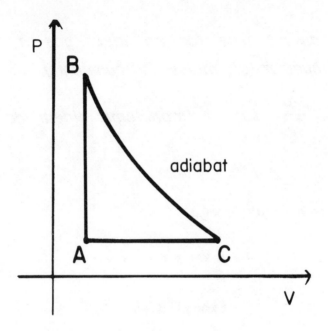

Figure 11.1

The area under a curve in the P-V plane is the work done. The area of a closed loop is the work done in a cyclic process.

Closed loop diagrams in the P-V plane represent a test-maker's dream and a test-taker's nightmare. The variety of problems that can be devised is endless. The main difficulty that you will have with such problems is that of organizing information. Even in the simplest case — a three-sided loop such as shown in Figure (11.1) — you must keep track of nine variables related by four equations:

$$
\begin{array}{llll}
A \rightarrow B & Q_{AB} & = & W_{AB} + \Delta U_{AB} \\
B \rightarrow C & Q_{BC} & = & W_{BC} + \Delta U_{BC} \\
C \rightarrow A & Q_{CA} & = & W_{CA} + \Delta U_{CA} \\
\text{add equations} & \text{net heat} & = & \text{net work} + 0
\end{array}
\tag{11.13}
$$

In Equations (11.13) we have used the fact that ΔU is always zero in a closed loop process. The reason is that the internal energy is a function of the <u>state</u> of the system. The state of a system is described by quantities like pressure, temperature, volume, etc. In a closed loop process, the system returns to its original state, so U remains the same, or "total" $\Delta U = 0$.

There are indeed nine variables related by four equations in Eqs. (11.13). It is not an accident that vertical lines appear among these equations. They suggest the following chart (which we shall modestly refer to as the Aaron-chart or A-chart):

	Q	W	$\triangle$U
A➔B	a	e	i
B➔C	b	f	j
C➔A	c	g	k
Totals	d	h	O

Table 11.1 The A-chart

The contents of Equations (11.13) are contained in Table 11.1 if we realize that the elements of the rows are related by the First Law of Thermodynamics.

$$Q = \Delta U + W$$

The column entries add to give the totals. You will find the A-chart an invaluable device for handling problems involving closed loop thermodynamic processes. Most problems of this type are straightforward if attacked systematically, but hopeless otherwise. Our chart automatically organizes information and forces a systematic solution. <u>It should be used even in the simplest of problems</u>!

One final note: Never forget the zero in the bottom right hand corner. Think of the advantage in a tic-tac-toe game if one square automatically belonged to you. The zero entry in the A-chart gives you such an advantage.

* * * * * * * * * * * * * *

Example 11-5

A thermodynamic system is taken through the cycle shown in Figure (11.1). Q = 20.0 J and the total work done is 15.0 J. Calculate the net heat added to the system during the process C ➔ A.

Solution: Fill in all the entries in the A-chart. (Remember that in an adiabatic process Q = 0, and in a constant volume process W = 0.)

	Q	W	△U
A→B	20 J	0	i
B→C	0	f	j
C→A	c	g	k
Totals	d	15 J	0

We now use the chart.

Bottom row:

$$d = 15 \text{ J} + 0$$

$$= 15 \text{ J}$$

First column:

$$d = 20 \text{ J} + 0 + C$$

$$c = -5 \text{ J}$$

The net heat added to the system during the process C→A is entry c, so $Q_{CA} = -5.0$ J, and the problem is finished!

<u>Our experience shows that most students cannot even begin doing such a problem without using a chart such as we suggest.</u>

11.4   The Second Law of Thermodynamics

The Second Law of Thermodynamics plays a secondary role in an introductory Physics course, but a primary role in our everyday lives. It states that

Heat cannot of itself pass
from one body to a hotter body.

Heat won't pass from a cooler to a hotter.

For you there are two main consequences of this law:

1. You will eventually die.

2. You cannot convert heat to work with 100% efficiency. For example, an engine operating between a hottest absolute temperature T_h and a coldest absolute temperature T_c has a <u>maximum</u> efficiency given by

$$e = \left(1 - \frac{T_c}{T_h}\right) \qquad (11.14)$$

This result is proven in most textbooks using the Carnot cycle.

Problems using Equation (11.14) are usually easy because they involve straight substitution. The only way to go wrong is to use degrees celsius rather then degrees kelvin. For example if you were told that T_c was the melting point of ice, we hope that you would not write

$$e = \left(1 - \frac{0^oC}{T_h}\right) = 1$$

We have presented two consequences of the Second Law of Thermodynamics. We shall not discuss consequence (1) any further. Consequence (2) follows from the fact that heat involves the kinetic energy associated with the chaotic motion of the atoms in a substance. To get useful work we must first (in some sense) organize this motion, and the organization itself uses energy. It's a shame that there is so much energy around us that we cannot use. Some individuals try to build engines with efficiencies as close as possible to that of Eq. (11.14); they are called engineers. Some individuals try to build engines with efficiencies greater than that of Eq. (11.14); they are called crackpots.

Energy energy everywhere

and all the tanks were dry.

Energy energy everywhere

why does everyone cry?

R. Aaron

12 ELECTROSTATICS

12.1   Introduction

Almost all phenomena that we encounter in everyday life are consequences of electric and magnetic forces. At the atomic and molecular level these forces are responsible for the structure of matter, so the mechanical forces we studied earlier are really electric effects --

Figure 12.1

In many ways the electromagnetic interaction is the most important subject in Physics. Electrical forces are responsible for chemical reactions, and therefore for life itself! Even optics is indirectly a study of the electromagnetic interaction, because light is an electromagnetic wave. Finally, most of the topics in Modern Physics (that you will study in Chapter 18) apply quantum mechanics to the electromagnetic interaction in order to explain atoms, molecules, and solids. So in a sense, the last third of an introductory Physics course is a study of electricity and magnetism and its applications.

Our material on electrostatics is arranged differently from that in most textbooks. We divide the chapter into two parts: Part A which treats point charges, and Part B which deals with continuous charge distributions.

Part A is easy, won't burden your minds, and will carry you through the 70-80% of the material on electrostatics which is most likely to appear on exams.

Part B isn't so easy, but we will guide you through the material.

PART A

12.2   Coulomb's Law

If two charges q_1 and q_2 are separated by a distance r, each charge experiences a force of magnitude

$$F = k_o \frac{|q_1 q_2|}{r^2} \quad (k_o = 9 \times 10^9 \frac{N \cdot m^2}{C^2}) \tag{12.1}$$

Charges of the same sign repel one another; charges of opposite sign attract one another.

In equation (12.1) both charges have equal billing. It is useful to rewrite this equation in a less symmetric way,

$$F = |q_2| E_1$$

$$E_1 = k_o \frac{|q_1|}{r^2} \tag{12.2}$$

where E_1 is defined as the magnitude of the electric field produced by the point charge q_1. Equation (12.2) is not symmetric with respect to the two charges -- there is a notion of cause and effect. The charge q_1 causes an electric field in space which affects other charges (like q_2).

Even though Equations (12.1) and (12.2) are really the same, try to think in terms of the electric field, rather than Coulomb's Law. Use Eq. (12.2) rather than Eq. (12.1) whenever possible. Do what we say -- it is for such brilliant advice that you purchased this book! The electric field is important for many reasons, but even at this stage it simplifies calculations as we show in Example 12-1.

150

＊ ＊ ＊ ＊ ＊ ＊ ＊ ＊ ＊ ＊ ＊ ＊ ＊ ＊

Example 12-1

A charge q_1 = +1.5 μC sits at rest at the origin.

(a) A second charge q_2 = -3.0 μC is placed at Point P, a distance 0.5 m from q_1. What is the force exerted by q_1 on q_2?

(b) Suppose the charge q_2 in part (a) is +2.2 μC. What is the force exerted by q_1 on q_2?

(c) Suppose the charge q_2 in part (a) is +7.0 μC. What is the force exerted by q_1 on q_2?

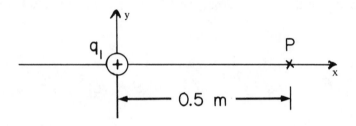

Figure 12.2

Solution: (a) First calculate the magnitude of the electric field E produced by q_1 at point P. Then multiply by q_2 to get the force.

$$E_1 = \frac{(9.0 \times 10^9 \text{ N-m}^2/\text{C}^2)(1.5 \times 10^{-6}\text{C})}{(0.5 \text{ m})^2}$$

$$= 5.40 \times 10^4 \text{ N/C}$$

$$F = E_1 |q_2| = (5.4 \times 10^4 \text{N/C})|-3 \times 10^{-6}\text{C}| = 0.16 \text{ N}$$

Since charges of opposite sign attract one another, the electrical force on q_2 acts to the left.

(b) We hope you were not planning to go through the same calculations as in part (a).

The electric field E at point P has not changed. Merely multiply the value of E_1 from part (a) by the new value of q_2 and you have your answer.

$$F = (5.4 \times 10^4 \, \text{N/C}) \, (2.2 \times 10^{-6} \, \text{C})$$

$$= 0.12 \, \text{N (to the right)}$$

(c)   By now you know to multiply the value of E_1 in part (a) by the new value of q_2 **Answer:** 0.38 N (to the right).

* * * * * * * * * * * * * *

The electric field is a vector: For point charges, it has magnitude given by Equation (12.2), and the direction depends on the sign of the charge. By definition, a positive charge produces an electric field which <u>points away from itself</u>, and a negative charge produces an electric field which <u>points toward itself</u>. Finally, a group of point charges q_1, q_2, q_3,... produce an electric field at any point in space which is the vector sum of the individual electric fields.

$$\boldsymbol{E} = \boldsymbol{E}_1 + \boldsymbol{E}_2 + \boldsymbol{E}_3 \tag{12.3}$$

A charge q brought to such a point feels a force

$$\boldsymbol{F} = q\boldsymbol{E} \tag{12.4}$$

* * * * * * * * * * * * * *

Example 12-2

Charges q_1 = +0.5 μC and q_2 = -1.3 μC are separated by 0.3 m. Where in space will a third charge q_3 = -7.5 μC feel no force.

Solution: The trick in this problem is to <u>think electric field</u>. Divide space into two parts; the first part is the <u>line joining the two charges</u>, and the second part is everywhere else (see Figure 12.3).

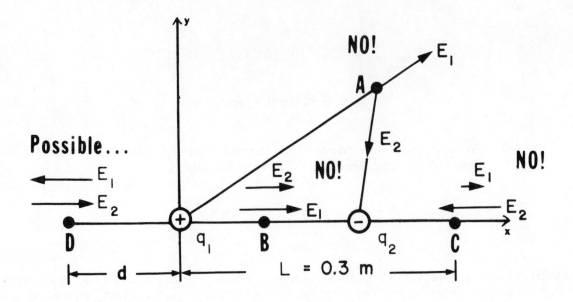

Figure 12.3

Consider Point A in Figure 12.3. This is an example of a point which is not on the line joining q_1 and q_2. The electric field contributions will be in different directions at such a point, and can never cancel. So there will always be a resultant electric field, and a third charge will always feel a net force.

It is also true that a third charge will feel a net force at a point such as B between the two charges on the line joining them. Because q_1 and q_2 have <u>opposite</u> sign, they contribute electric fields in the <u>same</u> direction at point B.

So we have narrowed it down to either Point C or Point D. The electric field cannot be zero at Point C because the larger charge is closer and will always dominate. Only at Point D can we play off the magnitudes of the charges against the distances from the charges, and hope that the electric fields cancel. The expression for the electric field at this point is

$$E_x = \frac{k_o |q_1|}{a^2} - \frac{k_o |q_2|}{(a+L)^2}$$

Setting $E_x = 0$, we solve for a.

$$|q_1| (a+L)^2 = |q_2| a^2$$

$$\sqrt{|q_1|} (a+L) = \pm \sqrt{|q_2|} \, a$$

$$\sqrt{0.5} (a+L) = \pm \sqrt{1.3} \, a$$

$$a = 0.490 \text{ m}, -0.115 \text{ m}$$

Only the positive solution makes sense.

* * * * * * * * * * * * * * *

Example 12-3

Charges q_1 = +4.5 μ C, q_2 = -2.0 μ C, and q_3 = -3.7 μ C are placed at the corners of a square as shown in Figure 12.4.

    (a) What is the electric field at point P?

    (b) Suppose a charge Q is placed at P. What electrostatic force acts on Q?

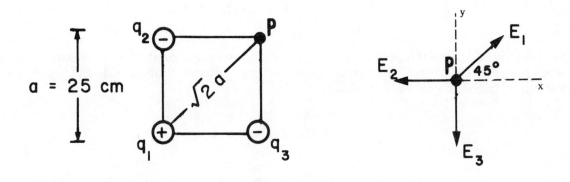

Figure 12.4

Solution: (a) Remember that the electric field is a vector. We are dealing with a simple vector addition problem. At Point P the contributions to the electric field are shown in Figure 12.4(b).

$$E_x^{net} = E_1 \cos 45^0 - E_2$$

$$E_y^{net} = E_1 \sin 45^0 - E_3$$

$$E_1 = \frac{k_0 q_1}{2a^2} = \frac{(9.0 \times 10^9 \text{ N-m}^2/\text{C}^2)(4.5 \times 10^{-6}\text{C})}{2(0.25 \text{ m})^2} = 3.24 \times 10^5 \text{ N/C}$$

$$E_2 = \frac{k_0 |q_2|}{a^2} = \frac{(9.0 \times 10^9 \text{ N-m}^2/\text{C}^2)(2 \times 10^{-6}\text{C})}{(0.25 \text{ m})^2} = 2.88 \times 10^5 \text{ N/C}$$

$$E_3 = \frac{k_0 |q_3|}{a^2} = \frac{(9.0 \times 10^9 \text{ N-m}^2/\text{C}^2)(3.7 \times 10^{-6}\text{C})}{(0.25 \text{ m})^2} = 5.33 \times 10^5 \text{ N/C}$$

Summing the x and y components we get

$$E_x^{net} = -0.59 \times 10^5 \, N/C$$

$$E_y^{net} = -3.04 \times 10^5 \, N/C$$

(b)   We get the force on any charge Q by multiplying the charge times the electric field calculated in part (a).

$$F_x^{net} = Q \, E_x^{net}$$

$$F_y^{net} = Q \, E_y^{net}$$

* * * * * * * * * * * * * *

The next example is important because it brings together concepts from Chapter 4 (forces) and the present chapter. We have pointed out before that professors love to give "combination" problems on examinations.

* * * * * * * * * * * * * *

Example 12-4

Two small balls with the same mass m carrying positive charges q_1 and q_2, are hung from a common point as shown in Figure 12.5.

Obtain expressions for T_1, T_2, θ_1 and θ_2, if the strings are of equal length, and the distance between the charges is r.

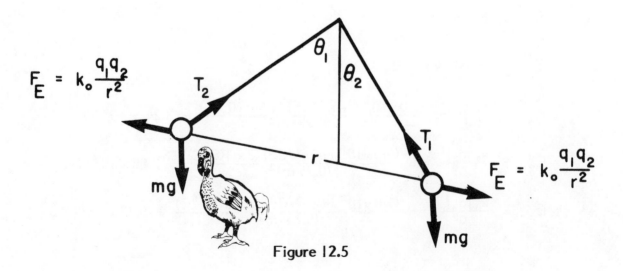

Figure 12.5

Solution: In this problem it is natural to think in terms of Coulomb's Law rather than the electric field. The main point here is that the electrical forces acting on the two balls depend on the product of the charges, so their magnitudes are the same. (Would this result suprise Newton?)

Thus Figure 12.5 is crazy! If the electrical (and gravitational forces) on each ball are of equal magnitude, and in the directions shown, why would one of the balls swing out to a larger angle than the other? The only alternative is that θ_1 and θ_2, and consequently T_1 and T_2 are equal. This result would follow from a detailed mathematical analysis, but such an analysis isn't necessary. Just use Figure 12.6 and the associated equations of equilibrium.

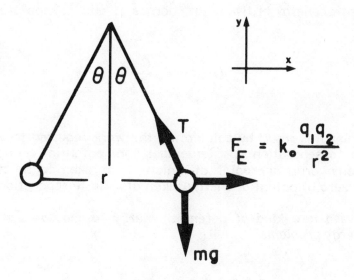

Figure 12.6

$$\sum F_x = \frac{k_0 q_1 q_2}{r^2} - T \sin \theta = 0$$

$$\sum F_y = T \cos \theta - mg = 0$$

Use methods of Example 4-10 to solve for T and θ

$$\tan \theta = \frac{k_0 q_1 q_2}{mgr^2}$$

$$T = \frac{mg}{\cos \theta}$$

* * * * * * * * * * * * * *

12.3 Electrical Potential and Potential Energy

The concepts of work and energy played a fundamental role in the subjects of mechanics and thermodynamics. They are equally important in studies of electricity. Just as masses have energy of position because of gravitational forces, charges have energy of position because of electrostatic forces. A group of positive charges confined in a small region of space has lots of potential energy. If allowed, they will spring apart and have the ability to do work. Because the electrostatic force is conservative we may define a potential energy function.

Remember that this is Part A of Chapter 12, in which we are dealing with point charges. The potential energy of two point charges q_1 and q_2 separated by a distance r is

$$U = k_o \frac{q_1 q_2}{r} \tag{12.5}$$

Your textbook obtains this result by calculating the work required to bring the charges from infinite separation to separation r. Remember that potential energy is always measured with respect to some arbitrary point at which it is taken to be zero. Equation (12.5) corresponds to the zero of potential energy taken at infinite separation of the charges.

You now have a new kind of potential energy to add to your collection, for use in conservation of energy problems.

* * * * * * * * * * * * * * *

Example 12-5

Nuclear forces are very strong, but only act over very short distances. On earth, the protons in a container of hydrogen gas never get close enough for nuclear reactions to take place. So you can carry a container of hydrogen gas, and not worry about it becoming a hydrogen bomb. The reason you don't have to worry, is that the electrical repulsion between the protons keep them out of range of the nuclear forces. At typical earthly temperatures, their kinetic energies are not large enough to sufficiently overcome their electrical repulsion.

Things are different on the sun. Temperatures near the center of the sun are about 12 million degrees kelvin. The average velocity of protons at this temperature is 3.0×10^6 m/s. Assume that two protons with this velocity start at infinite separation, and approach each other head on. What is their distance of closest approach?

Solution: Conservation of energy says that

$$2(\tfrac{1}{2} mv^2) = \frac{k_0 e^2}{a}$$

$$2\left[\tfrac{1}{2}(1.67 \times 10^{-27}\text{kg})(3 \times 10^6 \text{ m/s})^2\right] = \frac{(9.0 \times 10^9 \text{ N-m}^2/\text{C}^2)(1.6 \times 10^{-19}\text{C})^2}{a}$$

Since $1\text{ N} \equiv 1\text{ kg-m/s}^2$,

$$a = 1.53 \times 10^{-14}\text{m}$$

where a is the distance of closest approach. This distance is just within the range of nuclear forces, and the sun shines because nuclear reactions take place.

* * * * * * * * * * * * * *

Example 12-6

What is the potential energy of the system of four point charges at the corners of the square shown in Figure 12.7?

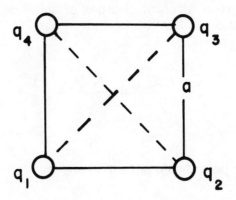

Figure 12.7

Solution: The potential energy of a charged pair is given by Equation (12.5). In the case of Figure 12.7 there are six such pairs, so there are six contributions to the total potential energy

$$U = \frac{k_0}{a}(q_1 q_2 + q_2 q_3 + q_3 q_4 + q_4 q_1)$$

$$+ \frac{k_0}{\sqrt{2}\, a}(q_1 q_3 + q_2 q_4)$$

By definition, the potential energy of a system of charges is the work required to construct it. Sometimes you will be asked to calculate the work required to build up a system of charges. Now you know that this is equivalent to calculating the potential energy of the charges.

* * * * * * * * * * * * * *

Electric Potential

Just as we introduced cause and effect into Coulomb's Law by defining the electric field, we can introduce the notion of cause and effect into Equation (12.5) by writing

$$U = q_2 V_1$$

$$V_1 = \frac{k_o q_1}{r} \quad (\text{has the algebraic sign of } q_1)$$

(12.6)

where V_1 is the potential produced by the point charge q_1; the charge q_1 <u>causes</u> a potential in space which <u>effects</u> other charges (like q_2).

Potential is measured in <u>volts</u>; 1 volt = 1 joule/coulomb

The potential produced by a group of point charges at some point in space is the algebraic sum of the individual contributions.

$$V = \frac{k_o q_1}{r_1} + \frac{k_o q_2}{r_2} + \ldots$$

(12.7)

A charge q brought to such a point has a potential energy

$$U = qV$$

(12.8)

Often you will be given the potentials V_1 and V_2 at two points in space, and asked for the gain or loss in potential energy when a charge q moves between these points. The answer is

$$\Delta U = q(V_2 - V_1)$$

(12.9)

* * * * * * * * * * * * * *

Example 12-7

A dental X-ray machine is basically a tube in which electrons start from rest at one end, pass through a difference in potential, and strike a target at the other end, producing X-rays. If the difference in potential between one end of the tube and the other is 80,000 V, what are the velocities of the electrons when they strike the target?

Solution: According to the law of conservation of energy, the increase in kinetic energy must be equal to the decrease in potential energy. Setting these two quantities equal we have

$$\frac{1}{2} m_e v_e^2 = e \, \Delta V$$

$$v_e = \sqrt{\frac{2e \, \Delta V}{m_e}} = \sqrt{\frac{2 \, (1.6 \times 10^{-19} C) \, (80,000 \, J/C)}{9.1 \times 10^{-31} \, kg}}$$

Since $1 \, J \equiv 1 \, N\text{-}m = 1 \, kg\text{-}m^2/s^2$, we get

$$v_e = 1.69 \times 10^8 \frac{m}{s}$$

This answer should make you pause for thought. The value $v = 1.69 \times 10^8$ m/s is so close to the velocity of light $c = 3.0 \times 10^8$ m/s that the kinetic energy is given only approximately by the expression $\frac{1}{2} mv^2$. In Chapter 18, in the section on special relativity, we show that the above answer is in error by about 10%.

* * * * * * * * * * * * * * *

Example 12-8

Find the electric potential at the midpoint of the line joining the two charges shown in Figure 12.8. What is the magnitude and direction of the electric field at this point (point P)?

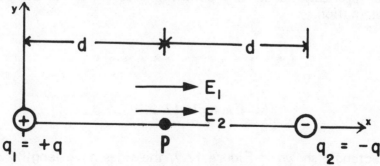

Figure 12.8

Solution: The potential at point P is the algebraic sum of each of the separate contributions. Since the two charges are equidistant from point P and equal in magnitude but opposite in sign, the sum of the individual contributions to the potential is zero.

$$V = \frac{k_0 q}{d} + \frac{k_0 (-q)}{d} = 0$$

The electric field at point P is <u>not</u> zero. As shown in Figure (12.8), the electric field contriubtions <u>add</u> to point P.

$$E = \frac{k_0 q}{d^2} + \frac{k_0 q}{d^2}$$

$$= \frac{2k_0 q}{d^2}$$

It is a common mistake, and a major sin to say that the electric field is zero because the potential is zero.

The numerical value of the potential cannot have physical significance because it depends upon the choice of origin. Remember that Equation (12.5) holds only because we have (arbitrarily) taken the potential energy to be zero at infinite separation of the charges. On the other hand it is physically significant that the electric field is zero at a point, since this means that no force will be exerted on any charge at that point. This assertion can be tested experimentally. As is shown in most calculus level textbooks, the electric field is determined from the derivative of the potential, not from the value of the potential at a point.

* * * * * * * * * * * * * *

The next example combines the main points of the previous two examples. If you can handle this, you can probably handle any potential problem (involving point charges) that appears on your examination.

* * * * * * * * * * * * * *

Example 12-9

In the rectangle shown in Figure 12.9, the sides have lengths 3.0 cm and 7.0 cm; $q_1 = 6.5$ µC and $q_2 = +4.0$ µC. How much work is required to move a third charge $q_3 = +2.5$ µC from B to A?

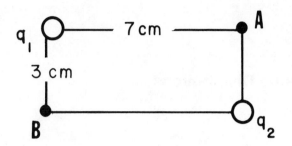

Figure 12.9

Solution: This is a simple problem and can appear on either a calculus level or non-calculus level Physics examination. Don't make it hard by trying to integrate, or doing something else crazy!

Just perform the following three simple steps:

(i) Calculate the potential at A.

$$V_A = \frac{k_o q_1}{(0.07m)} + \frac{k_o q_2}{(0.03m)}$$

$$= \frac{(9.0 \times 10^9 \text{ N-m}^2/\text{C}^2)(6.5 \times 10^{-6}\text{C})}{(0.07 \text{ m})} + \frac{(9.0 \times 10^9 \text{ N-m}^2/\text{C}^2)(4.0 \times 10^{-6}\text{C})}{(0.03 \text{ m})}$$

$$= 2.04 \times 10^6 \text{ V}$$

(ii) Calculate the potential at B.

$$V_B = \frac{k_o q_1}{(0.03m)} + \frac{k_o q_2}{(0.07m)}$$

$$= 2.46 \times 10^6 \text{ V}$$

(iii) Subtract V_A from V_B, and multiply the result by q_3 (recall that 1 V $\equiv$ 1 J/C).

$$W = (2.5 \times 10^{-6}\text{C})(2.04 \times 10^6 \text{ J/C} - 2.46 \times 10^6 \text{ J/C})$$
$$= -1.05 \text{ J}$$

If you did not treat the powers of ten by hand, but rather you plugged numbers like 2.5×10^{-6}(exponential and all) into your calculator, step to the front of the line and receive an extra large

* * * * * * * * * * * * * *

PART B

12.4   Continuous Charge Distributions

Until this section we have considered only point charges as the sources of the electrostatic force. We shall now study cases where the source of electrostatic force is a continuous charge distribution (no graininess), which occupies an extended region of space. There are two ways to treat such charge distributions.

1.   Integration: Since Coulomb's Law for point charges is the only tool at our disposal, we can resort to the same approach suggested in Section 8.2. The trick is to break up the continuous charge distributions into tiny elements which behave like point charges, and use integration rather than simple addition. We shall not discuss this approach further. Such problems rarely appear on introductory Physics exams, and if they do, they are very similar to examples in the text.

2.   Gauss' Law: Gauss' Law is a reformulation of Coulomb's law developed (by guess whom?) in the 19th century. It involves advanced mathematics, but it is easy to use for the problems that you will encounter. If a problem on continuous charge distributions appears on an exam, it will probably be a "Gauss' Law problem".

12.5   Gauss' Law

In all standard texts, Gauss' Law is developed in terms of the concept of electric field lines. The direction of the electric field E as one travels through space can be represented by continuous lines. The electric field is tangent to these lines at any point in space. Not only are they useful in visualizing field directions, but they can be used to display the magnitude of E in each region of space. The lines are closer together when E is large, and farther apart when E is small.

Following Faraday, one can take the concept of electric field lines or electric flux more seriously, and actually develop numerical relations between the number of lines and the electric field. For constant electric field E and flat surface A, the number of electric field lines that pass through the surface (see Figure 12.10) is defined to be

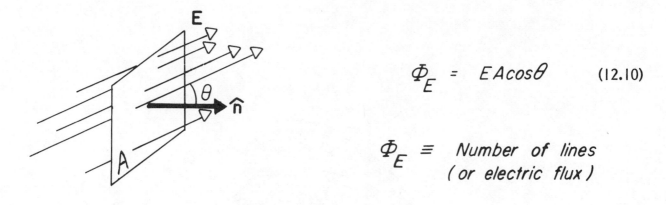

$$\Phi_E = EA\cos\theta \qquad (12.10)$$

$$\Phi_E \equiv \ \text{Number of lines} \\ \text{(or electric flux)}$$

Figure 12.10 $\hat{n}$ is a unit vector normal to surface.

where θ is the angle between the electric field and the normal to the surface. This definition leads to the result that the number of lines emerging from a point charge is

$$\Phi_E = \frac{q}{\epsilon_o} \quad \text{(electric flux of a point charge)} \qquad (12.11)$$

Putting Equations (12.10) and (12.11) together, and using the superposition principle

$$E = E_1 + E_2 + \ldots.$$

leads to Gauss' Law which states that the total number of electric field lines Φ_E emerging from any <u>closed</u> surface that we can <u>imagine</u> is

$$\Phi_E = \sum_{\substack{\text{Gaussian} \\ \text{surface}}} E\Delta A\cos\theta = \frac{q_{inside}}{\epsilon_o} \qquad (12.12)$$

The sum in Equation (12.12) takes place over our <u>imaginary</u> or <u>Gaussian</u> surface, and q_{inside} is the total charge <u>inside</u> the Gaussian surface. In Figure 12.11 we show a strange shaped surface immersed in a non-uniform electric field.

If you think that you can perform the sum indicated by Equation (12.12) for Figure 12.11, you have delusions of grandeur. Fortunately we will only encounter simple surfaces on which E is constant, and cos θ is ± 1.0 or 0. For these cases Equation (12.12) reduces to something simple, such as

$$\sum E\Delta A\cos\theta \longrightarrow EA \qquad (12.13)$$

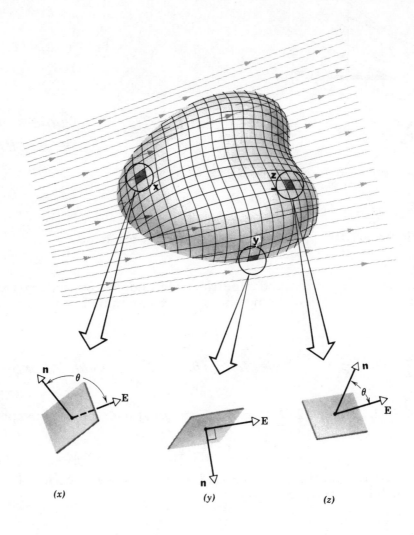

Figure 12.11 We focus on three contributions to the sum in Eq. (12.12). **n** is a unit vector perpendicular to the element surface area ΔA

    In case (x), cos θ is negative
    In case (y), cos θ is zero
    In case (z), cos θ is positive

If the surface encloses zero charge the sum in Eq. (12.12) must be zero. The zero sum comes about because some of the E ΔA cos θ are positive, while others are negative.

The amazing thing is that the sum "knows" to be zero for any shape closed surface in any electrostatic field.

Gauss' Law is only useful for calculating electric fields when Equation (12.13) can be used. This happens for very few situations with a high degree of symmetry. In fact, being able to handle the five cases shown in Figure 12.12 is sufficient for almost any introductory Physics course.

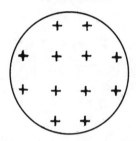

(a) Uniform spherical ball of charge.

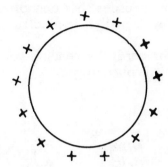

(b) Uniform spherical shell of change.

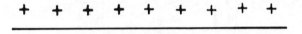

(c) Uniform, infinite, line of charge.

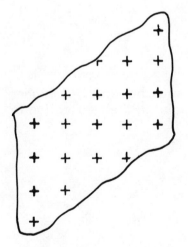

(d) Uniform, infinite, sheet of charge.

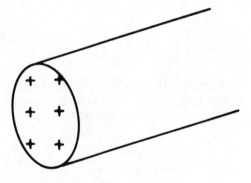

(e) Uniform, infinite, cylinder of charge.

Figure 12.12

For the cases in Figure 12.12, calculating the electric field using Coulomb's Law is much more laborious and complicated than using Gauss' Law. However, there are only five cases, and it would be easier to memorize these five results than introduce the complex machinery of Gauss' Law. Why introduce Gauss' Law? The answer is that Physicists use Gauss' Law not only for calculational purposes, but because it gives insights into the nature of electrical forces that cannot be obtained by other means. For example, it can be shown using Gauss' Law that in electrostatics, charges always reside on the surface of a conductor.

After carefully reading your textbook, and this book, many of you will still feel that Gauss' Law is pure magic.

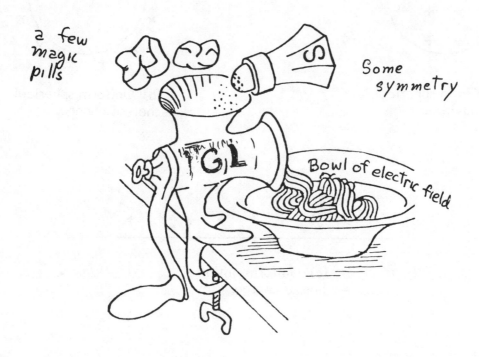

Students view of Gauss' Law

It is perfectly normal to feel this way. Gauss' Law is a peek into the world of higher mathematics. Only a few students have developed enough at this point in their education to really appreciate the material.

* * * * * * * * * * * * * *

Example 12-10

In Figure 12.13 we show a spherically symmetric charge distribution of total charge Q and radius R, with a Gaussian surface of radius r (greater than R) surrounding it. By the density of dots we have tried to indicate that the density of charge varies with distance from the center, but not with lattitude or longitude angles. Find an expression for the electric field when r > R.

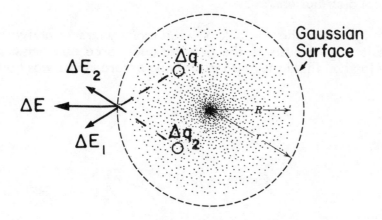

Figure 12.13 You choose the Gaussian surface so that it is symmetric with respect to the charge distribution and passes through the point at which you want the electric field.

Solution: If you don't want to lose points on an exam, you should write words like the following before using Gauss' Law: "By symmetry this is a case where E is everywhere perpendicular to the Gaussian surface, and has constant magnitude over the surface. Therefore I may use the simplified version of Gauss' Law."

Your professor will be impressed by the phrase "by symmetry". But in case you do not quite understand what it means, we shall explain it. Every element of charge Δq in the charge distribution has a partner as shown in Figure 12.13. These Δq's contribute electric fields whose components parallel to the surface cancel. The magnitude of E is the same everywhere on the Gaussian surface because the charge distribution looks the same from everywhere on the Gaussian surface.

We now use the simplified version of Gauss' Law given in Equation (12.13). Remember that A is the area of the Gaussian surface, and q_{inside} is the charge <u>inside</u> the surface.

$$EA = \frac{q_{inside}}{\varepsilon_o}$$

$$E(4\pi r^2) = \frac{Q}{\varepsilon_o}$$

$$E = \frac{Q}{4\pi\varepsilon_o r^2}$$

This example gives a clue to the power of Gauss' Law. We have proven that <u>any</u> spherically symmetric charge distribution appears as a point charge when one is outside the charge distribution. This is true for cases (a) and (b) in Figure 12.12. Since they are special cases of a spherical distribution.

You might note that it took Newton many years to arrive at a similar result for the case of the gravitational force and mass. Of course Newton had the disadvantage of dying before Gauss was born.

* * * * * * * * * * * * * * *

Example 12-11

In Figure 12.14 we show a uniform (thin) spherical shell of total charge Q.

(a) What is the electric field at point P?

(b) What is the potential at point P?

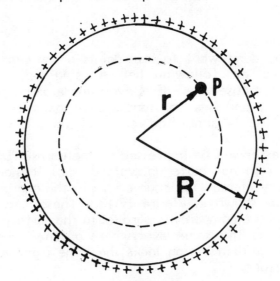

Figure 12.14

Solution: (a) Say the correct words as explained in Example 12-10. Then use the simplified version of Gauss' Law. In this case no charge is enclosed by the Gaussian surface [see Figure (12.14)].

$$EA = \frac{q_{inside}}{\varepsilon_o}$$
$$EA = 0$$
$$E = 0 \qquad \text{(since } A = 4\pi r^2 \text{ is not zero)}$$

(b) We are dealing with a spherically symmetric charge distribution, so as proven in Example 12-10, the electric field is that of a point charge Q everywhere outside the spherical shell of charge. Thus the potential is that of a point charge right up to the surface of the shell.

$$E = \frac{Q}{4\pi\epsilon_o r} \quad \text{(from } r = \infty \text{ to } r = R\text{)}$$

so

$$V = \frac{Q}{4\pi\epsilon_o r} \quad \text{(from } r = \infty \text{ to } r = R\text{)}$$

In part (a) we proved that the electric field is zero inside the shell, so no work is required to move a charge from the surface to point P. Therefore the potential is the same at point P as it is at the surface:

$$E = 0 \quad \text{(r less than R)}$$

$$V = \frac{Q}{4\pi\epsilon_o r} \quad \text{(r less than R)}$$

The potential is not zero because the electric field is zero! We have already shown you in Example 12-8 that the electric field is not zero because the potential is zero.

* * * * * * * * * * * * *

We have not considered the uniform solid ball of charge, the infinite uniformly charged wire, or the infinite uniformly charged sheet of charge. The reason is that although they may appear on exams, there are no tricks involved, and they are covered carefully in most textbooks. *Just be sure to memorize the formulas for the surface areas of spheres and cylinders.* We bring you down to earth with a final example which is really an old fashioned force problem. It is a minor variation of the standard statics problem that professors love to ask on exams.

* * * * * * * * * * * * * *

Example 12-12

A particle of mass m with charge Q is suspended by a thin thread in a parallel plate capacitor as shown in Figure (12.15). If m = 5 g, Q = +2.5 μC, and θ = 30°, find the tension in the string and the surface charge density on the capacitor plate.

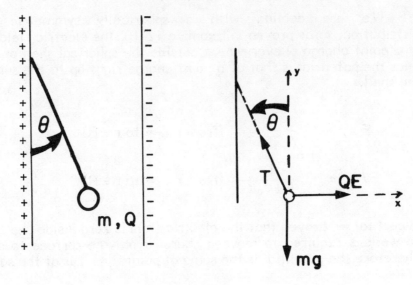

Figure 12.15

Solution: You should memorize the fact that the electric field between two capacitor plates is

$$E = \frac{\sigma}{\varepsilon_0} \quad (\sigma = \text{charge per unit area})$$

The force diagram now gives the following equations of equilibrium:

$$\sum F_x = Q\left(\frac{\sigma}{\varepsilon_0}\right) - T \sin\theta = 0$$
$$\sum F_y = T \cos\theta - mg = 0$$

We now solve these equations for $\tan\theta$ by dividing (as in Example 12-4).

$$\tan\theta = \frac{Q\sigma}{mg\varepsilon_0}$$
$$\sigma = \frac{mg\,\varepsilon_0\,\tan\theta}{Q}$$

* * * * * * * * * * * * * * * *

13 ELECTRICAL CIRCUITS

13.1 Introduction

Electrical circuits is rightfully a topic for either applied Physics or engineering. It is a little out of place in an ordinary Physics lecture course. Circuit elements and electrical measuring devices seem very abstract when one reads about them, but they do not seem at all abstract when one "tinkers" with them in the laboratory. The circuit problems in an elementary Physics text are usually cut and dry, so in this chapter we give relatively few examples, and spend more time trying to make the concepts appear less abstract. However, nothing that we can do can take the place of a few well spent laboratory sessions.

In order to maintain a current in a wire, we need a steady source of electrical energy. Such a source of electrical energy is called a seat of emf, or simply an emf. Batteries and generators are examples of seats of emf. In a battery it is chemical energy that is converted into electrical energy, and in a generator it is mechanical energy. We shall use the symbol $\mathcal{E}$ for emf --

$$\mathcal{E} = \frac{\Delta W}{\Delta q} \tag{13.1}$$

where ΔW is the electrical energy given to a charge Δq when it passes through the seat of emf.

The simplest electrical circuit consists of a battery (an emf) connected to a resistor by perfectly conducting wires. A resistor is a circuit element that dissipates electrical energy in the form of heat. In Figure 13.1 we show such a circuit; we have used the standard symbol for resistance.

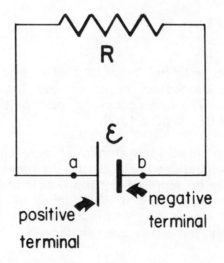

Figure 13.1

Such diagrams as Fig. 13.1 are very schematic; for example, the resistor will often represent devices such as a toaster or a light bulb.

The source of emf in Fig. 13.1 creates a difference in potential between points a and b by separating positive and negative charges. A current of positive charges (according to electrical engineering conventions) then flows from the positive terminal to the negative terminal of the emf. When the charges were separated they were given potential energy. When they have traversed the circuit and neutralized one another, all that potential energy is gone. What happened to it?

The answer is that the electrical potential energy of the moving charges was converted to heat energy in the resistor. Conservation of Energy tells us that the energy dissipated in the resistor must be exactly equal to the energy provided by the emf. Another way of stating this conclusion is that

> The algebraic sum of the potential differences (or voltages) across all the circuit elements (including resistors, capacitors, emfs, whatever...) in a closed loop of an electrical circuit must be zero.

Kirchhoff's First Law

Another important property of electrical circuits follows from the principle of conservation of charge. At a junction of wires, the current entering the junction must equal the current leaving the junction. This idea is embodied in Kirchhoff's second law:

> At a junction of wires, the current entering the junction, must equal the current leaving the junction.

Kirchhoff's Second Law

Kirchhoff's two laws as stated previously are the fundamental equations of circuit theory.

13.2 Resistance and Ohm's Law

In this chapter we shall deal with two types of electrical circuit elements -- resistors and capacitors. To explain the concept of resistance we turn to an analogy with blood flowing in the coronary arteries. The heart plays the part of the emf, and the blood plays the part of the electrical current (see Figure 13.2).

When blood flows in a blood vessel (or water in a pipe), friction between the blood and the walls of the vessel and friction between neighboring elements of blood, tends to stop the flow. So energy must continually be supplied (by the heart) to keep the blood flowing. In Figure 13.2(a) a lesion has developed due to "hardening of the arteries" providing an additional resistance, and the patient is sick. In Figure 13.2(b) a second lesion has developed in series with the first one, and the patient is even sicker. We see that resistances in series add to one another.

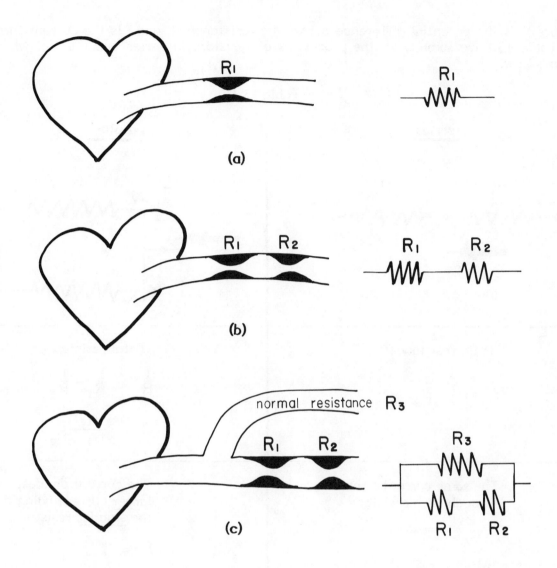

Figure 13.2

In Figure 13.2(c) a surgeon has relieved the patient's problem by grafting a new blood vessel in parallel with the first one. According to the figure, the damaged vessel still helps a bit; it takes some load off the new graft. So the total resistance of a set of parallel resistors is less than that of any single resistor in the set (see Example 13-1).

In electrical circuits, the atomic structure of the conductor provides a resistance which impedes the flow of electrons, and the emf provides the energy which keeps it going. In an introductory Physics course we study series and parallel combinations of resistors. We are only interested in resistances which obey Ohm's Law.

$$V = IR \qquad (13.2)$$

where V is the potential difference across the resistance R, and I is the current through R. In Table 13.1 we summarize the properties of resistors in series and parallel which obey Ohm's Law.

Table 13.1

<u>series</u>	<u>parallel</u>
R₁ and R₂ in series from a to c to b, with current I	R₁ and R₂ in parallel between a and b, with current I dividing into I₁ and I₂

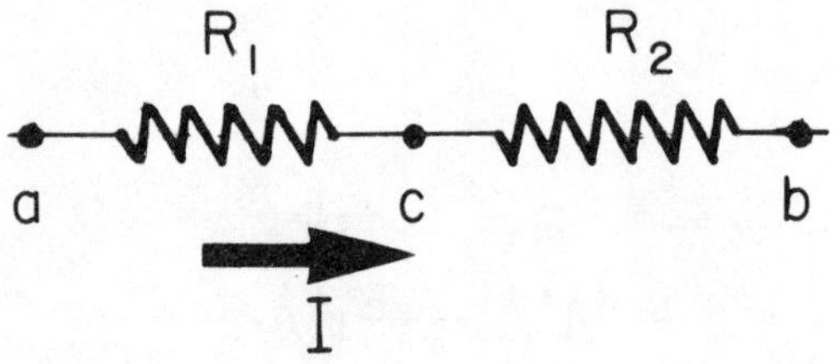

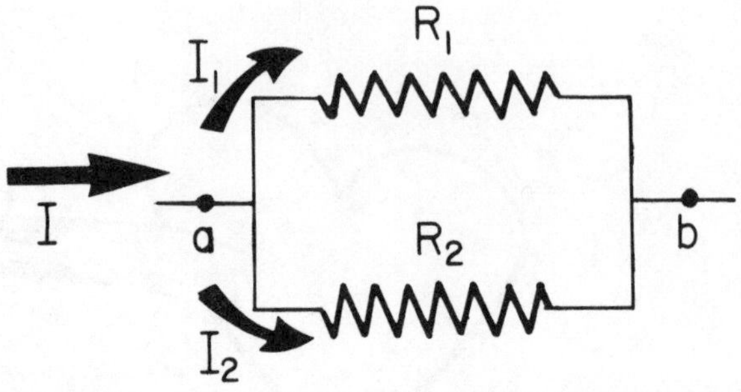

series	parallel
Total resistance: $$R = R_1 + R_2$$	Total resistance: $$\frac{1}{R} = \frac{1}{R_1} + \frac{1}{R_2}$$ $$R = \frac{R_1 R_2}{R_1 + R_2}$$
The same current I passes through each resistor.	The current divides, with the <u>larger</u> current going through the <u>smaller</u> resistor. $$I = I_1 + I_2$$
The voltage divides: $$V_{ab} = V_{ac} + V_{cb}$$ The larger voltage drop occurs across the larger resistor.	The voltage is the same across circuit elements in parallel.

Each resistor obeys Ohm's Law

$$V = IR$$

The power dissipated in such a resistor is

$$P = IV$$
$$= I^2 R$$
$$= V^2/R$$

In Examples 13-1 and 13-2 we apply the rules developed in Table 13.1.

* * * * * * * * * * * * * *

Example 13-1

Two resistance R_1 and R_2 are in parallel. Show that the total resistance R is less than either R_1 or R_2.

Solution: You will never see a problem like this on an examination. But study it anyway! It will sharpen your algebra skills, and help you avoid one of the most horrible of all algebraic errors (which we shall mention shortly). The proof proceeds in the following way:

$$\frac{1}{R} = \frac{1}{R_1} + \frac{1}{R_2}$$

$$= \frac{R_1 + R_2}{R_1 R_2}$$

We now invert the equation, and then divide numerator and denominator on the right hand side by R_1.

$$R = \frac{R_1 R_2}{R_1 + R_2} = \frac{(R_1 R_2)/R_1}{(R_1 + R_2)/R_1}$$

$$R = \frac{R_2}{1 + R_2/R_1}$$

Since the denominator is greater than one, we have proven that R is less than R_2. In the same way one can show that R is less than R_1.

The Most Horrible Error

Once again we wavered between insulting your intelligence and warning you. We have decided to warn you. We have seen many students write the following:

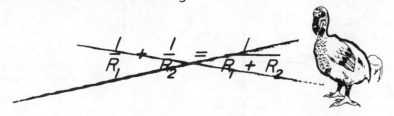

Never do this; bring to a common denominator and proceed as we have done earlier in this example. This "dodo" error is bad because it puts your exam grader in a particularly nasty frame of mind. Not only will the grader be unkindly disposed toward you, but he will take out his inner frustrations on the next ten papers that he grades. Even worse, one of these students will make the same error. You have the opportunity to break this terrible chain of events -- do it!

* * * * * * * * * * * * * *

Example 13-2

In Figure 13.3 we show a battery connected to a system of identical light bulbs and switches. Describe the relative brightness of the three bulbs when

(a) Switch S_1 is closed; S_2 and S_3 are open.

(b) Switches S_1 and S_2 are closed; S_3 is open.

(c) Switches S_1, S_2, and S_3 are all closed.

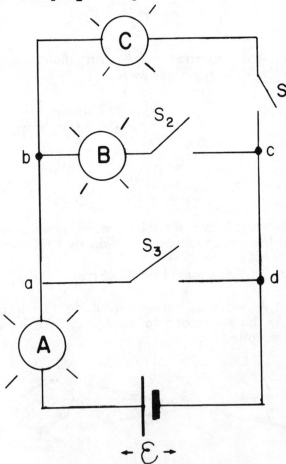

Figure 13.3

Solution: This is a practical problem. You will even learn what is meant by a short circuit.

The main concept in this problem is that the brightness is proportional to the power dissipated. From Table 13.1 we have

$$P = I^2 R \quad or \quad V^2/R \tag{13.3}$$

Since the light bulbs are identical, they have the same resistance. Thus Equation (13.3) tells us that the bulb with the largest voltage across it will be the brightest.

(a) If you follow the current from the positive terminal of the battery back to the negative terminal, you will see that it passes through light bulbs A and C; light bulb B is not in the circuit because switch S_2 is open. For this case, Figure 13.3 can be replaced by the schematic circuit diagram of Fig. 13.4.

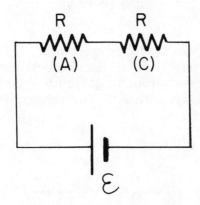

Figure 13.4

So light bulbs A and C are lit (with equal brightness), and bulb B is dark.

(b) In this case light bulbs B and C form a parallel combination in series with light bulb A. All three light bulbs are lit. To understand what is happening let's examine the currents through the three resistors. If a current I leaves the emf, it will pass through resistors A, and then divide so that only ½ I passes through each of the resistors B and C. Since $P = I^2R$, we may conclude that light bulb A (which receives the most current) is brightest. Since light bulbs B and C receive the same current (½ I) they are equally bright, but less bright than A.

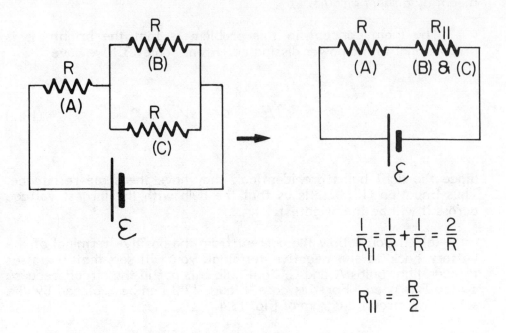

$$\frac{1}{R_{||}} = \frac{1}{R} + \frac{1}{R} = \frac{2}{R}$$

$$R_{||} = \frac{R}{2}$$

Figure 13.5

(c) When all three switches are closed, light bulbs B and C are "short circuited". There is no resistance between points a and d, so after the current passes through light bulb A and arrives at junction a, it will <u>all</u> pass through wire <u>ad</u>. The schematic diagram in Figure 13.6 illustrates this effect.

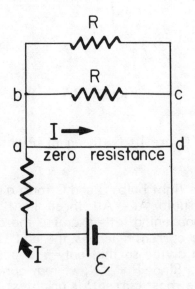

Figure 13.6

13.3 Capacitors in Series and Parallel

A capacitor stores charge; it consists of two conductors which carry equal and opposite charge. The capacitance C tells the amount of charge Q that can be stored at voltage V.

$$C = \frac{Q}{V}$$

(13.4)

In Table 13.2 we outline the properties of series and parallel combinations of capacitors. It is useful to note that the formulas for capacitors are opposite to those for resistance. The formula for capacitors in series is like that for resistors in parallel, and vice versa.

Table 13.2

series	parallel
Total capacitance: $\frac{1}{C} = \frac{1}{C_1} + \frac{1}{C_2}$ $C = \frac{C_1 C_2}{C_1 + C_2}$	Total capacitance: $C = C_1 + C_2$
The same charge Q is on each capacitor.	The charge divides. The capacitor with the greater capacitance has the greater charge. $Q = Q_1 + Q_2$

Table 13.2

(Con't.)

.

The voltage divides: $$V_{ab} = V_{ac} + V_{cb}$$	The voltage is the same across circuit elements in parallel.

The energy in a capacitor is

$$U = \frac{1}{2}QV$$
$$= \frac{1}{2}CV^2$$
$$= \frac{1}{2}\frac{Q^2}{C}$$

The subject of capacitors in series and parallel is a highly technical one, and of little interest to one not building electrical devices. The main Physics concept involved is that capacitors store electrostatic energy, and most examination questions test understanding of this idea. Typically, two capacitors with different charges and different voltages are connected together, and you are asked to find the final energy. The important fact that you must remember is that for situations like this, the final energy is always <u>less</u> than the initial energy. When the capacitors are connected sparking takes place, and resistance in the wires opposes the redistribution of charge. There is no way of avoiding these energy losses.

* * * * * * * * * * * * * * *

Example 13-3

A capacitor with C = 12 μF is charged to a potential difference of 5,000 V. The charging battery is then removed and the capacitor is connected as in Figure 13.7 to an uncharged capacitor with C = 4.5 μF.

(a) If the switch S is thrown, what is the final potential difference across the combination?

(b) What is the stored energy before and after the switch is thrown?

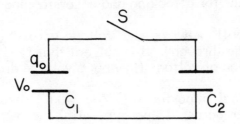

Figure 13.7

Solution: (a) When the capacitors are connected, we have a parallel combination of capacitors (the potential difference across each is the same, but the charges on each are different). The original charge q is now shared by the two capacitors; thus

$$q_0 = q_1 + q_2$$

Applying the relation $q = CV$ to each of these terms yields

$$C_1 V_0 = C_1 V + C_2 V$$

or

$$V = V_0 \frac{C_1}{C_1 + C_2}$$

$$= (5,000 \text{ V}) \left(\frac{12}{12 + 4.5} \right) \frac{10^{-6}}{10^{-6}} \text{ V}$$

$$= 3,636 \text{ V}$$

(b)   The initial stored energy is

$$U = \frac{1}{2} C_1 V_0^2$$

$$= \frac{1}{2} (12 \times 10^{-6} \text{ F}) (5,000 \text{ V})^2$$

$$U = 150 \text{ J}$$

The final stored energy is

$$U = \frac{1}{2} C_1 V^2 + \frac{1}{2} C_2 V^2$$

$$= \left[\frac{1}{2} (12 \times 10^{-6}) (3,636)^2 + \frac{1}{2} (4.5 \times 10^{-6}) (3,636)^2 \right] \text{ J}$$

$$= 109 \text{ J}$$

Note that the final energy is less than the initial energy (as promised).

We give you some final tips for attacking and understanding circuit problems.

1. Language is important! Always say "voltage across" and "current through". We have heard students say "voltage through". It is current that <u>flows through</u> a wire like water through pipes. There is no notion of voltage flowing; a voltage drop means an energy loss.

These ideas have practical importance. You would measure a voltage drop in a resistor by placing a voltmeter across (in parallel with) the resistor. The current through the resistor is measured by placing an ammeter in series with the resistor -- you want the same current to flow through the ammeter that flows through the resistor.

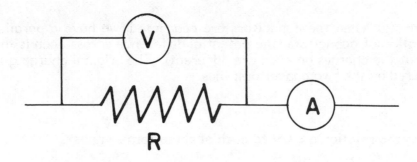

Figure 13.8

Ideally, the resistance of the voltmeter in Figure 13.8 should be infinite. In this case no current would flow through the voltmeter, and the ammeter would read the current flowing through the resistor. Similarly, the ideal ammeter would have zero resistance, or else it would disturb the circuit (by having a voltage drop across it). In practice, the best we can do is build voltmeters with very large resistances, and ammeters with very small resistances.

2. Don't be fooled by drawings! All three diagrams in Figure 13.9 describe the same situation -- a resistor and capacitor in parallel with an emf $\mathcal{E}$.

Note that the resistors and capacitors in Figure 13.10 <u>are neither in series nor parallel</u>. These are real two loop circuits which must be analyzed using Kirchhoff's Laws. We choose not to treat such circuits in this chapter.

13.4 RC Circuits

In Figure 13.11 we show simple one loop circuits which contain resistance and capacitance (hence the name RC). Until this section our currents were constant as functions of time. In this section we deal with currents that vary with time, but are always in one directions (direct currents). For <u>direct currents</u>, a capacitor acts as a break in the circuit, and <u>eventually</u>, no current flows.

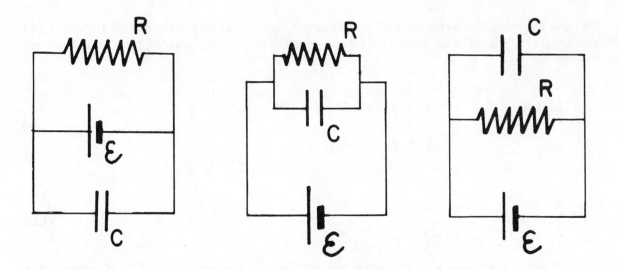

Figure 13.9

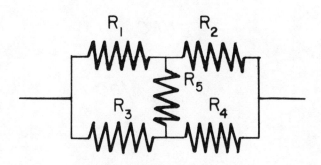

Figure 13.10

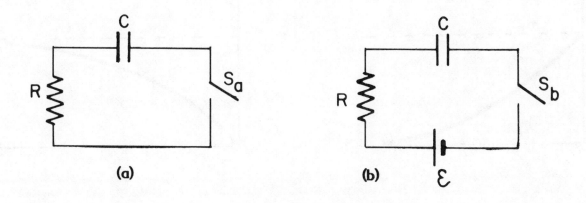

Figure 13.11

We now examine the behaviors of the charges, currents, and voltages in Figure 13.11. For these circuits, Kirchhoff's first law (the sum of the voltages around the circuit is zero) takes the form

From Fig. 13.11a:

$$iR + \frac{q}{C} = 0 \quad (i = \frac{dq}{dt})$$

$$R\frac{dq}{dt} + \frac{q}{C} = 0$$

(13.5)

From Fig. 13.11b:

$$iR + \frac{q}{C} = \mathcal{E}$$

$$R\frac{dq}{dt} + \frac{q}{C} = \mathcal{E}$$

(13.6)

Thus we see that Kirchhoff's law yields two differential equations. In an introductory Physics course you are almost never required to solve differential equations. In this case it is sufficient to know that the solutions of Equations (13.5) and (13.6) involve the magic function exp(-t/RC) in one of the following combinations

$$e^{-t/RC}$$

(13.7a)

$$1 - e^{-t/RC}$$

(13.7b)

These objects are plotted as functions of time in Figure 13.12. The quantity RC is called the <u>time</u> <u>constant</u> τ, and controls the behavior of RC circuits at large times.

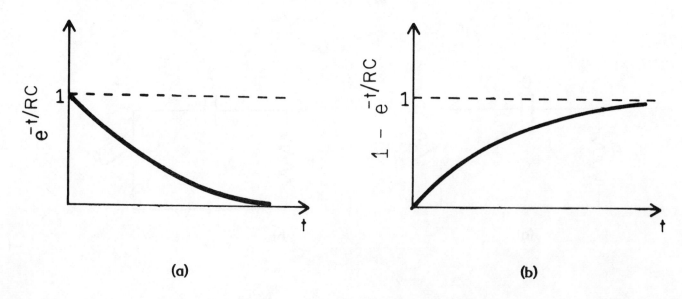

(a)   (b)

Figure 13.12

Figures 13.12(a) and 13.12(b) suggest how to use Equations (13.7a) and (13.7b). The function exp(-t/RC) describes physical quantities which have finite value at t = 0, and fall to zero as t → ∞. On the other hand, [1 - exp(-t/RC)] describes physical quantities that are zero at t = 0, and approach a finite constant as t → ∞.

Most introductory Physics problems involving RC circuits can be solved using the relations in Equation (13.7), together with the simple properties of resistors and capacitors that you have already learned.

* * * * * * * * * * * * * * *

Example 13-4

At t = 0 Herb (R = 40,000 Ω) grabs two ends of a 20 µF capacitor holding a charge of 4,000 µC (See Figure 13.13).

(a) What is the time constant of the effective circuit?

(b) What current is passing through Herb when t = 1.2 s?

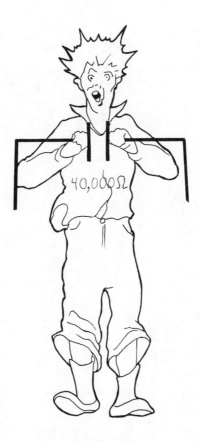

Figure 13.13

Solution: (a) First replace Figure 13.13 by a schematic diagram. For this example, Figure 13.11a is the correct diagram. Herb is the resistor (40,000 Ω), and grabbing the capacitor corresponds to closing the switch s_a. The time constant of the circuit is

$$RC = (40,000 \ \Omega)(20 \times 10^{-6} F)$$

$$= 0.8 \ s$$

(b)   The charge falls to zero from its initial value of 4,000 μC, so we use Equation (13.7a) to get

$$q = (4,000 \times 10^{-6} C) \, e^{-t/(0.8 \ s)}$$

$$i = \frac{dq}{dt} = -\frac{(4,000 \times 10^{-6} C)}{(0.8 \ s)} e^{-t/(0.8 \ s)}$$

$$= -(0.005 \ A) \, e^{-t/(0.8 \ s)}$$

When t = 1.2 s

$$i = -(0.005 \ A) \, e^{-1.5}$$

$$= -1.12 \times 10^{-3} A$$

* * * * * * * * * * * * * *

14 THE MAGNETIC FIELD

14.1 Introduction

We shall assume that you are familiar with the previous material on electrostatics and electrical circuits, and tell you the following story:

> In Figure 14.1 we show a long straight wire carrying a steady current I = 1,000 A, and the difference in potential between points A and B is 1,000,000,000 V (one <u>billion</u> volts).   A small positive charge q is placed a distance r from the wire.

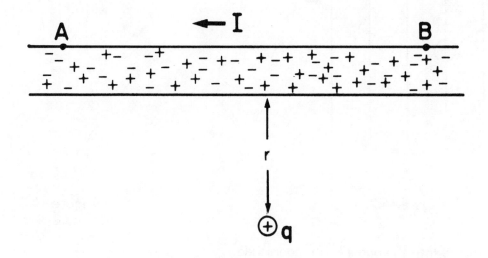

Figure 14.1

We notice a curious phenomena. If q is at rest it feels no force. If it moves to the left it feels a force upward, and if it moves to the right it feels a force downward. Your reaction to this story should be --

> (a) With a thousand amps and a billion volts, the wire must have a ton of charge on it! No wonder charge q feels a force.

(b) But I thought that a current carrying wire was neutral...

(c) The title of this chapter is "The Magnetic Field," so the charge must be feeling a magnetic force.

Reactions (b) and (c) are acceptable. Reaction (a) is wrong. A wire carrying current is neutral. Many charges are moving, but they are moving through a background of charges of the opposite sign. Any small volume of the wire contains zero net charge, so according to electrostatics the charge q should not feel an electrical force. We are indeed observing a new phenomenon: the magnetic force.

14.2   Charges in a Magnetic Field

When two charges are moving, we observe a new kind of force (in addition to the electrical force); its magnitude and direction depend on the velocities of the charges in an unusual way. In Figure 14.2 we show a charge q moving with velocity **v** at an angle θ to a magnetic field **B**.

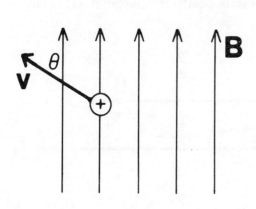

Figure 14.2

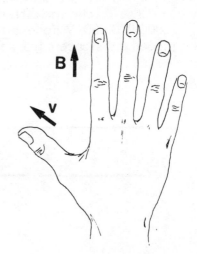

Figure 14.3

It feels a force <u>into the paper</u> with magnitude

$$F = qvB\sin\theta \qquad (14.1)$$

The direction of this force is obtained from your favorite right hand rule. We suggest the following one:

Keep your left hand behind your back!

Stare at the fingers of your Right hand -- wiggle them a bit. They look like lines of magnetic field, don't they? From now on think of your fingers as lines of magnetic field. Now look at your thumb -- it describes a charge moving with velocity **v**. From now on whenever you see the thumb of your right hand think of a positive charge moving in the direction of your thumb points. If you have been properly conditioned by the previous discussion, Figures 14.2 and 14.3 should convey the same meaning. Now push your hand ahead; that's the direction of the magnetic force -- into the paper.

In engineering (and some pre-med) texts, the magnetic force is written in terms of the vector cross product.

$$F = qv \times B$$

$$v \times B \equiv cross \; product \; of \; v \; and \; B$$

(14.2)

In advanced Physics and engineering courses, the cross product is used extensively. However, in most introductory Physics courses it is just a short hand code. For example **v** x**B** is a concise way of writing a vector with magnitude $vB \sin \theta$, and direction given by a right hand rule.

Circular Motion

Since the magnetic force is perpendicular to the velocity, it can only change its direction. Thus a charged particle entering a region of magnetic field will move in a curved path with constant speed. In Figure (14.4) we show a charge q moving perpendicular to a constant magnetic field **B**. In this case, the charge moves in a circle.

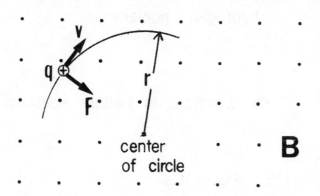

Figure 14.4 The magnetic field **B** is out of the paper.

Newton's Second Law gives the relation

$$F = ma$$

$$qvB = m\left(\frac{v^2}{r}\right)$$

(14.3)

$$r = \frac{mv}{qB}$$

Typical exam questions usually combine Equation (14.3) with some ideas from electrostatics.

* * * * * * * * * * * * * * *

Example 14-1

A deuteron and an α -particle are accelerated through the same difference of potential, and enter a region of uniform magnetic field. The particles are moving at right angles to the field B.

In Figure 14.5 we show a schematic representation of a deuteron and an α-particle. [For this problem you may take the masses of the proton and neutron to be equal.]

(a) Compare their kinetic energies.

(b) If the radius of the deuteron's circular path is 12.0 cm, what is the radius of the α-particle's path?

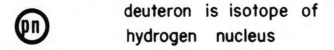 deuteron is isotope of hydrogen nucleus

 α – particle is helium nucleus

Figure 14.5

Solution: (a) In this problem it is important not to substitute numbers too quickly.

deuteron: $\quad \frac{1}{2}(2m)v_d^2 = e\,\Delta V$

α-particle: $\quad \frac{1}{2}(4m)v_\alpha^2 = 2e\,\Delta V$

(b)   We may obtain the velocities of the deuteron and α-particle from knowledge of their kinetic energies.

$$v_d = \sqrt{\frac{e\,\Delta V}{m}}$$ (different masses and different charges compensate to give the same velocities)

$$v_\alpha = \sqrt{\frac{e\,\Delta V}{m}}$$

Equation (14.3) now gives the radii.

$$r_d = \frac{2m\sqrt{e\,\Delta V/m}}{e\,B} \quad ; \quad r_\alpha = \frac{4m\sqrt{e\,\Delta V/m}}{2e\,B}$$

$$r_d = r_\alpha \;!!! \;\text{(different masses and different charges}$$
compensate to give same radii)

* * * * * * * * * * * * * *

14.3   Sources of the Magnetic Field

We have seen that moving charges experience magnetic forces. Also moving charges exert magnetic forces; they are the sources of magnetic fields (just as stationary charges are the sources of electrostatic fields). In electrostatics you used Coulomb's Law and Gauss' Law to calculate electric fields from known static charge distributions. The corresponding laws for calculating magnetic fields produced by moving charges or current distributions are the Biot-Savart Law and Ampere's Law.   The bad news is that these laws are relatively complicated, and we cannot improve upon the discussions in most textbooks, and therefore will not discuss them further in this book. The good news is that a single result from all this material will carry you through most final exams (not necessarily midterms and quizzes). You must know that a long straight wire produces a magnetic field that circles the wire and has magnitude

$$B = \frac{\mu_o I}{2\pi r} \qquad (\text{circles the wire}) \qquad (14.4)$$

$$\mu_o = 4\pi \times 10^{-7}\,\frac{T-m}{A}$$

where r is the distance from the wire. The direction of the field is given by the right hand rule demonstrated in Figure 14.6.

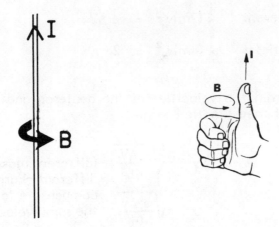

Figure 14.6

Once again your fingers (which curl around the wire) are the lines of magnetic field, and your thumb points in the direction of the moving charges. Remember, this is a right hand rule. If you ever use your left hand instead of your right hand, replace the **th** in thumb by a D!

One more important fact -- lines of electric field begin and end on charges; lines of magnetic field have no beginning and ending.

14.4   Magnetic Forces on a Current

Suppose a wire of length L carries a current I perpendicular to a uniform magnetic field B (see Figure 14.7).

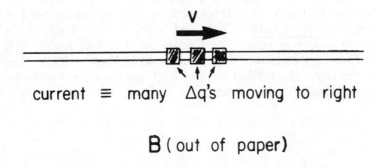

current ≡ many Δq's moving to right

B (out of paper)

Figure 14.7

The William Effect

In Figure 14.7 we have many tiny charges Δq, each feeling a tiny force ΔF given by Equation (14.1).

$$\Delta F = \Delta q v B \qquad (down) \qquad\qquad (14.5)$$

Using Equation (14.5) we can derive a formula for the total force on the wire. [We have simplified matters by having the wire perpendicular to the magnetic field. If the wire were at an angle θ to the field, a factor of $\sin \theta$ would appear in Equation (14.5).]

Suppose that Δq moves a distance ΔL in a time Δt; then

$$v = \frac{\Delta L}{\Delta t}$$

$$I = \frac{\Delta q}{\Delta t}$$

Now substitute v into ΔF and slide the Δt under the Δq.

$$\Delta F = \Delta q \frac{\Delta L}{\Delta t} B$$

$$= \frac{\Delta q}{\Delta t} \Delta L B$$

$$\Delta F = I \Delta L B$$

Summing all the tiny ΔF's gives the total force F.

$$F = BIL \qquad (down) \qquad\qquad (14.6)$$

BIL is the nickname for William; hence the title of this subsection. Our hope is that this disgustingly bad joke will imprint the very important Equation (14.6) indelibly on your brain.

* * * * * * * * * * * * * *

Example 14-2

A wire 60 cm long, and with a mass of 10 gm is suspended by a pair of flexible leads in a magnetic field of 0.40 T. What are the magnitude and direction of the current required to remove the tension in the supporting leads?

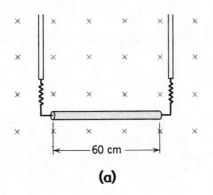

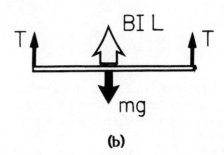

(a) **(b)**

Figure 14.8

Solution: This is an old fashioned force problem, so first construct the force diagram shown in Figure 14.8(b). The current must be to the <u>right</u> in order for the magnetic force to be <u>upward</u>. The conditions for equilibrium are

$$F_y = 2T + BIL - mg = 0$$

Setting the tensions equal to zero, we now solve for the current I. You want your current in amps, so you must convert centimeters and grams to meters and kilograms.

$$I = \frac{mg}{BL}$$

$$= \frac{(0.01 \text{ kg}) (9.8 \text{ m/s}^2)}{(0.40 \text{ T}) (0.50 \text{ m})}$$

$$I = 0.49 \text{ A}$$

This is a relatively straightforward problem; the only twist is that one of the forces is a magnetic force. The most likely error is to mix units.

* * * * * * * * * * * * * * *

Example 14-3

Figure 14.9 shows a long wire carrying a current of 50 A. The rectangular loop carries a current of 30 A. Calculate the resultant force acting on the loop if a = 1.5 cm, b = 12.0 cm, and L = 40 cm.

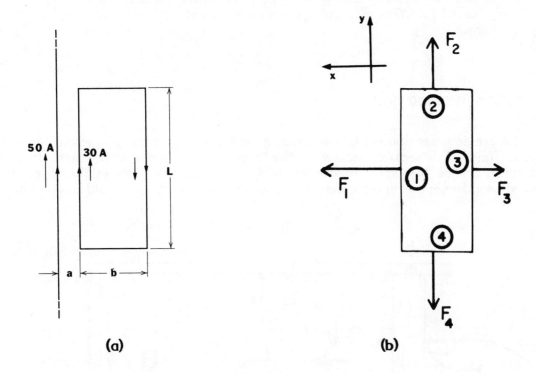

(a)　　　　　　　　　　　　**(b)**

Figure 14-9

Solution: In Figure 14.9(b) we show the four forces acting on the four wires of the rectangular loop. F_2 and F_4 are equal in magnitude (by symmetry), but opposite in direction (because the currents are in opposite directions). Thus the net force on the loop is to the left.

$$F_x^{net} = F_1 - F_3$$

$$= \frac{\mu_o (50 \text{ A})}{2\pi(0.015 \text{ m})} (30 \text{ A})(0.4 \text{ m}) - \frac{\mu_o (50 \text{ A})}{2\pi(0.135 \text{ m})} (30 \text{ A})(0.4 \text{ m})$$

Since $\mu_o = 4\pi\times10^{-7}$ T-m/A

$$F_x^{net} = \left[\frac{2(50)(30)(0.4)}{0.015} - \frac{2(50)(30)(0.4)}{0.135} \right] \times 10^{-7} \text{ T-A-m}$$

$$= \left[\frac{1}{0.015} - \frac{1}{0.135} \right] 2(50)(30)(0.4) \times 10^{-7} \text{ N}$$

$$= 7.11 \times 10^{-3} \text{ N}$$

As we shall show in Example 14-5, it requires calculus to calculate F_2 and F_4. You are lucky that they cancel and you may neglect them.

* * * * * * * * * * * * * *

14.5 Motion Induced emf

In the previous section we learned that a wire carrying a current would feel a force in a magnetic field. This section deals with the "opposite" or "inverse" process. If we exert a force to push a wire through a magnetic field, a current will flow in the wire. In Figure 14.10 we show a finger pushing a conducting rod through a magnetic field B with velocity v.

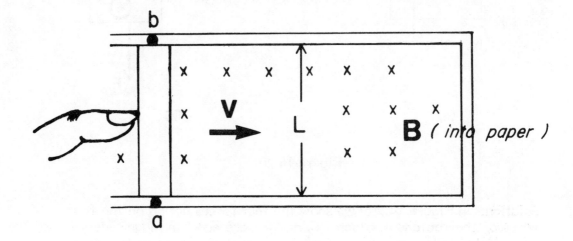

Figure 14.10 B is constant.

The rod is always in contact with a conducting track, so we have a closed circuit. Thus as we push the rod forward a current I will move upward, and then around the circuit. We have created a simple-minded electrical generator. It may be hard to believe at this time, but our civilization runs on such devices. Even the fuel that runs most electrical generators is similar to the fuel that runs the finger; namely, organic material from dead plant and animal remains.

Let us return to Figure 14.10. Because we are forcing the rod through the magnetic field B with velocity v, each tiny element of conduction charge Δq feels a tiny force ΔF given by Equation (14.1).

$$\Delta F = \Delta q v B \quad (up) \tag{14.7}$$

In Chapter 12 we learned that if there is a force per unit charge at a point, an electric field exists at that point. Applying this idea to Equation (14.7), we see that there is an electric field in the moving rod of Figure 14.10, given by

$$E = \frac{\Delta F}{\Delta q}$$

$$= vB \tag{14.8}$$

The electric field acts through a distance L, so the wire is a seat of emf:

$$emf \equiv (electric\ field) \times (displacement)$$

$$\mathcal{E} = BLv \tag{14.9}$$

It must be understood that the properties of the electric field defined by Equation (14.8) are much different from those of the electrostatic field that we studied in Chapter 12. Most important, the work done in moving a charge from point a to point b in Figure 14.10 depends upon the path taken. For example, if our path is outside the wire, no work is done on the charge! So the "electrodynamic" field of Equation (14.8) is not conservative.

We will now give you an important example, that has a long history of appearing on examinations. In fact we would say that this example is one of the three most predictable examination questions. The other two are: a trajectory problem (in kinematics) and a ballistic pendulum problem (for momentum and energy conservation).

* * * * * * * * * * * * * *

Example 14-4

The conducting rod (of length 2.0 m) in Figure 14.10 is pushed at a constant speed of 1.5 m/s, and the magnetic field has magnitude 3 T. Because of the motion of the rod, there is a current of 10 amperes in the circuit.

(a) What is the induced emf in the circuit?

(b) What is the force with which the finger is pushing the rod?

(c) What is the resistance of the circuit abcd?

(d) What is mechanical power produced by the finger?

(e) What power is dissipated in the resistance of the circuit?

Solution: (a) We get the emf directly from Equation (14.9)

$$\mathcal{E} = BLv$$

$$= (3\ T)\ (2\ m)\ (1.5\ m/s)$$

$$= 9V$$

(b)   There is a force $\underline{BIL}$ on the wire. Because the wire is moving at constant velocity (no acceleration) this force must be equal to the force exerted by the finger:

$$F = BIL \qquad \text{(to the left)}$$

$$= (3\ T)\ (10\ A)\ (2\ m)$$

$$= 60\ N$$

(c)   Ohm's Law gives the resistance:

$$R = \frac{\mathcal{E}}{I}$$

$$= \frac{9}{10}\ \Omega$$

$$= 0.9\ \Omega$$

(d)   Just plug in the formula power = force × velocity.

$$P = (60\ N)\ (1.5\ m/s)$$

$$= 90\ W$$

(e)   Use Table 13.1. If your answer is different from that in part (d), you're in trouble!

$$P = I^2 R$$

$$= (100\ A)\ (0.9\ \Omega)$$

$$= 90\ W$$

* * * * * * * * * * * * * * *

14.6   A Calculus Example

Problems involving the "real" use of calculus rarely appear on the examinations of an introductory Physics course.   Still you should know when the calculus is required for solving problems.   This knowledge gives you important insight into the subject at hand, and puts you "one up" on the students who don't care.   In Figure 14.9, the magnetic field produced by the long wire varies over wires #2 and #4, so you cannot use <u>BIL</u> for the whole wires.   But you can use it for little pieces of wire of length L, and this fact should signal <u>integration</u> in your mind.

* * * * * * * * * * * * * *

Example 14-5

Obtain expressions for the forces on wires #2 and #4 in Figure 14.9. I_1 flows in the long straight wire and I_2 flows in the loop.

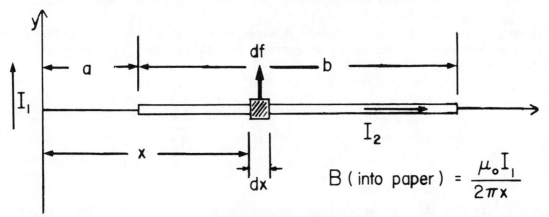

$$B \text{ (into paper)} = \frac{\mu_o I_1}{2 \pi x}$$

Figure 14.11 Wire #2

Solution: Consider wire #2.   Break it up into tiny elements of length <u>dx</u> as shown in Figure 14.11.   Each element feels a tiny force df upward given by "BIL":

$$df = BI_2 \, dx = \frac{\mu_o I_1 I_2}{2 \pi x} \, dx$$

We obtain the total force F by integrating (adding up) all the df's as the <u>dummy</u> <u>variable</u> x runs from a to a+b.

$$F = \frac{\mu_o I_1 I_2}{2 \pi} \int_a^{a+b} \frac{dx}{x} = \frac{\mu_o I_1 I_2}{2 \pi} \ln\left(\frac{a+b}{a}\right) \quad \textbf{(up)}$$

The force on wire #4 has the same magnitude and is directed downward.

15 FARADAY'S LAW, SELF INDUCTANCE

15.1   Introduction

It is almost a magical fact that when the magnetic lines of force threading a circuit change (either increase or decrease), a current begins to flow in the circuit.   This phenomenon is described by Faraday's Law,

$$\mathcal{E} = -\frac{d\Phi}{dt} \qquad \left(\mathcal{E}_{av} = -\frac{\Delta\Phi}{\Delta t} \right) \qquad\qquad (15.1)$$

In Equation (15.1) $\mathcal{E}$ is the emf or voltage around the circuit, and Φ is the magnetic flux threading the circuit. We shall now consider three aspects of Faraday's Law:

1.   emf around a circuit --

The emf or voltage <u>around</u> a circuit is a new concept. In electrostatics we learned that if a charged particle <u>gains</u> energy in moving between two points, the two points are at different potentials. But in electrostatics, if a charge moves in a closed path, back to its starting point, no energy is gained or lost, because <u>the electrostatic field is conservative</u>.

The potential difference associated with Faraday's Law is different. A changing magnetic flux produces an electric field which is not conservative. Each time a charge moves around a closed loop, this new kind of electric field can do work, and the charge can gain or lose energy.

2.   Magnetic flux --

In an introductory Physics course, problems involving Faraday's Law deal mostly with magnetic fields which are constant in <u>space</u>, and flat surfaces. Figure 15.1 shows constant magnetic field lines passing through a flat surface at an angle θ.

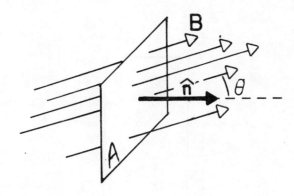

Figure 15.1 $\hat{n}$ is a unit vector normal to surface.

The magnetic flux ϕ threading the path in Figure 15.1 is defined to be

$$\Phi = AB\cos\theta \qquad (15.2)$$

The flux ϕ through a closed path will change and produce an emf, if any one of the three quantities B, A, or cos θ change.

3. The minus sign (Lenz's Law) --

We shall <u>not</u> treat the minus sign in Equation (15.1) as a strictly mathematical quantity. Rather we treat it as a symbol which reminds us of Lenz's Law ....

15.2 Lenz's Law

Lenz's Law tells us the direction of the emf (the direction that a current of positive charges will flow) when the flux through a circuit is changing. The current will flow in a direction that opposes the conditions that gave rise to it. In other words, if the flux threading a circuit is <u>decreasing</u>, the current will flow so that its own magnetic field tends to <u>increase</u> the flux threading the circuit.

Lenz's Law is one of those cases where one picture is worth a thousand words, so consider Example 15-1.

Example 15-1

In what direction does the current flow in each of the five situations shown in Figure 15.2?

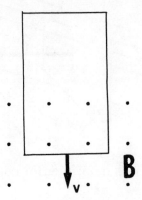

(a) Loop falls through magnetic field.

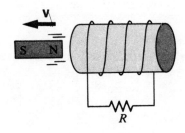

(b) Bar magnet moves to the left.

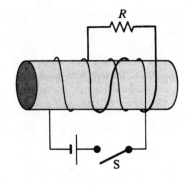

(c) Close switch S.

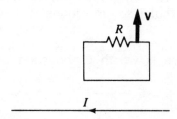

(d) Loop moves perpendicular to wire.

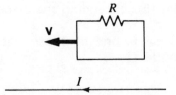

(e) Loop moves parallel to wire.

Figure 15.2

Solution: (a) In Figure 15.2(a) the flux threading the loop (out of the paper) increases as the rectangular loop falls; current will flow clockwise, producing a magnetic field which passes through the loop, into the paper.

(b) As the bar magnet recedes from the solenoid, the number of magnetic field lines (pointing to the left) passing through the loops of wire decreases. A current will flow up (between points a and b) in order to give more lines to the left.

(c) When the switch is closed, the lines of magnetic field pointing to the left through the solenoid will increase. Trying to prevent this increase, a current will flow in the second circuit, from left to right across the resistor.

(d) If the rectangular loop moves away from the long straight wire, the flux <u>into</u> the paper through the rectangular loop is <u>decreasing</u>. Thus a current flows <u>clockwise</u> in the loop, trying to maintain the flux.

(e) This is the dodo question. If the loop moves parallel to the long straight wire, the flux threading the loop remains constant, and no current flows.

* * * * * * * * * * * * * * *

15.3 Application of Faraday's Law

In the introduction we stated Faraday's Law,

$$\mathcal{E}_{av} = -\frac{\Delta \Phi}{\Delta t} \tag{15.1}$$

where

$$\Phi = AB\cos\theta \tag{15.2}$$

for constant magnetic field B at an angle θ to a flat surface area A (see Fig. 15.1). We further pointed out that an emf is produced if any one of the three dependent variables in Equation (15.2) changed. In this section we give one example for each case.

Example 15-2

Derive Equation (14.9) for the emf in Figure 15.3 using Faraday's Law, rather than the methods of Chapter 14.

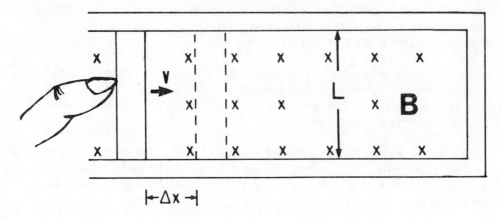

Figure 15.3

Solution: This is an example in which the area is changing. From Figure 15.3 we see that in a time Δt, the moving wire has traveled a distance Δx; thus the area has decreased by an amount $L\,\Delta x$. Faraday's Law now gives the emf --

$$\mathcal{E} = -\frac{\Delta \Phi}{\Delta t}$$

$$\Delta \Phi = B\,\Delta A$$

$$\Delta A = -L\,\Delta x \qquad \text{(minus sign means decreasing)}$$

$$\mathcal{E} = BL\,\frac{\Delta x}{\Delta t} \qquad \left(\frac{\Delta x}{\Delta t} \equiv v\right)$$

$$\mathcal{E} = BLv \qquad \text{(Equation 14.9)}$$

* * * * * * * * * * * * * *

Example 15-3

A circular conductor of radius a = 0.5 m is rotated about an axis AC as shown in Figure 15.4, at a constant rate of 5 revolutions per second. A uniform magnetic field is directed into the paper, and has magnitude 5.0 T.

(a) What is the average emf produced in one quarter of a revolution? In one complete revolution?

(b) Calculate the maximum emf produced in the conductor.

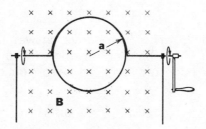

Figure 15.4

Solution: (a) In this example, it is the angle between B and the normal to the plane (that contains the conducting loop), which is changing. At any time t the flux threading the coil is given by

$$\Phi = BA \cos \theta \quad (\theta = \omega t = 2\pi ft = \frac{2\pi}{T}t)$$

$$= BA \cos\left(\frac{2\pi}{T}t\right)$$

In one quarter of a period the flux Φ changes from its maximum value of BA to 0 (or _vice_ _versa_).

$t = 0:$ $\qquad \Phi = BA \cos 0^o = BA$

$t = \frac{T}{4}:$ $\qquad \Phi = BA \cos \frac{\pi}{2} = 0$

$$\mathcal{E}_{av} = \frac{\Delta \Phi}{\Delta t} = \frac{BA-0}{T/4}$$

$$= \frac{4BA}{T} \quad \text{(magnitude of average emf)}$$

(b) The maximum emf is obtained from Faraday's Law. At any time t we have

$$\mathcal{E} = -\frac{d\Phi}{dt} = -\frac{d}{dt}(BA \cos \omega t)$$

$$= \omega BA \sin \omega t$$

(If you know no calculus memorize the above result.) The maximum value of $\mathcal{E}$ is the quantity in front of the sine function, since sin (ωt) never exceeds unity.

$$\mathcal{E}_{max} = \omega BA$$

$$= \frac{BA}{T}$$

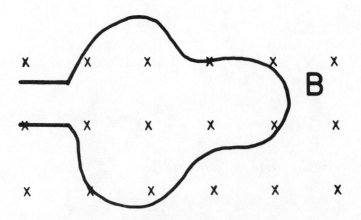

You were not concentrating. The above answer is wrong! How could the <u>maximum</u> emf (BA/T) be smaller than the <u>average</u> emf (4BA/T)? We have purposely made an error to keep you awake. The correct answer is

$$\mathcal{E}_{max} = \omega BA \qquad \omega = \frac{2\pi}{T}$$

$$= \frac{2\pi BA}{T} \qquad (2\pi \text{ is larger than 4 })$$

* * * * * * * * * * * * * * *

Example 15-4

A wire loop of area A is placed in a uniform magnetic field B as shown in Figure 15.5. The magnetic field is perpendicular to the plane of the loop. It changes with time according to the equation

$$B = bt^3$$

(a) Write an expression for the emf induced in the loop.

(b) In what direction does the current flow?

Figure 15.5 B into paper.

Solution: (a) Use Faraday's Law to get the emf.

$$\Phi = BA$$

$$= bAt$$

$$\mathcal{E} = -\frac{d\Phi}{dt} = -3bAt$$

(b)   The magnetic field into the paper is increasing, so by Lenz's Law a current will flow <u>counterclockwise</u> in order to decrease B.

* * * * * * * * * * * * * * *

15.4   Self Inductance—LC Circuits

While we use derivatives in this section, no calculus is required for the examples, which often appear on "non-calculus" exams.

In Figure 15.6 we show a solenoid carrying current i.

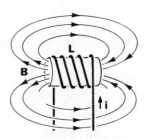

Figure 15.6

If i is constant, B is constant, and there is no potential difference between points a and b. On the other hand, if i starts changing, B starts changing, and according to Lenz's Law, the loops in the solenoid send current in a direction which opposes the change in i.   Thus the solenoid is acting like a source of emf, and a potential difference appears between points a and b.   This <u>induced</u> <u>emf</u> is given by the formula

$$\mathcal{E}_{ab} = L\frac{di}{dt} \tag{15.3}$$

where L is called the <u>self inductance</u> of the coil.   Until now you have considered resistors and capacitors as circuit elements.   Now you have a third circuit element for your repertoire -- it is called an <u>inductor</u> (something that has an inductance L).

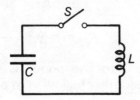

Figure 15.7

In Figure 15.7 we show a circuit which contains a charged capacitor and an inductor (an LC circuit). When the switch s is closed, a current begins to flow, and the net charge on the capacitor plates oscillates (changes sign) as a function of time. For Figure 15.7, Kirchhoff's law for voltages says that

$$L\frac{di}{dt} + \frac{q}{C} = 0 \tag{15.4}$$

Substituting i = dq/dt into Equation (15.4) yields the differential equation

$$\frac{d^2q}{dt^2} = -\omega^2 q$$

$$\omega = \frac{1}{\sqrt{LC}} \tag{15.5}$$

From Chapter 9 we know that this differential equation describes simple harmonic motion. Thus the charge q is oscillating with angular frequency ω.

* * * * * * * * * * * * * * *

Example 15.5

Suppose that the oscillation frequency of the LC circuit in Figure 15.8(a) is 24,000 Hz. What will the frequency become when two more capacitors identical to the first are added in the manner shown in Figure 15.8(b).

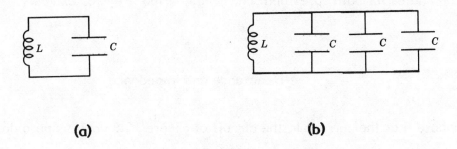

Figure 15.8

Solution: In Figure 15.8(a) the frequency is

$$f_a = \frac{\omega}{2\pi} = \frac{1}{2\pi\sqrt{LC}}$$

In Figure 15.8(b) the capacitance has tripled (since the capacitors are in parallel), so·

$$f_b = \frac{1}{2\pi\sqrt{3LC}}$$

Thus

$$f_b = f_a/\sqrt{3} \qquad = 24{,}000 \text{ Hz}/\sqrt{3}$$

$$= 13{,}856 \text{ Hz}$$

* * * * * * * * * * * * * * *

15.5 Alternating Current Circuits (RLC)

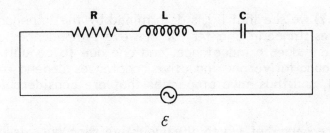

Figure 15.9

In this section we shall discuss two simple aspects of alternating current theory, leaving the hard stuff for your textbook and your instructor. The subjects we stress are strong candidates for both "pre-med" and "engineering" Physics exams.

Reactance and Impedance

Suppose that the current in the circuit of Figure 15.9 varies sinusoidally according to

$$i = i_{max} \sin \omega t \qquad (15.6)$$

Then a relation similar to Ohm's Law connects the **maximum values** of $\mathcal{E}$ and i:

$$i_{max} = \frac{\mathcal{E}_{max}}{Z} \qquad (15.7)$$

$$Z = \sqrt{R^2 + (X_L - X_C)^2} \equiv impedance$$

$$X_L = \omega L \equiv inductive\ reactance$$

$$X_C = \frac{1}{\omega C} \equiv capacitative\ reactance$$

From Equation (15.7) we see that i_{max} is determined by the impedance Z, which plays a role similar to that of resistance in Chapter 13. However, impedance has three components: one due to resistance, one due to inductance, and one due to capacitance. It is important to realize that the capacitive and inductive reactances depend on the frequencies of the alternating current, and thus have properties that are considerably different from that of resistance.

There is a frequency at which the inductive reactance is equal to the capacitative reactance, and the combined effects cancel each other in the expression for impedance. This frequency is called the <u>natural</u> or <u>resonance</u> frequency of the circuit.

$$X_L = X_C$$

$$\omega L = \frac{1}{\omega C}$$

$$\omega^2 = \frac{1}{LC}$$

$$f_o = \frac{\omega}{2\pi} = \frac{1}{2\pi\sqrt{LC}} \quad \left(\begin{array}{l}\text{resonant or} \\ \text{natural frequency}\end{array}\right) \qquad (15.8)$$

At frequency f_o, the impedance Z is minimum (equal to R) and maximum current flows. *Another property of the resonant frequency is that it usually finds its way into final exams!*

* * * * * * * * * * * * * *

Example 15.6

An LC combination in an AM radio tuner uses an inductor with L = 0.1 mH and a variable air capacitor such as shown in Figure 15.10. What capacitance will tune the radio (have maximum current flow) at a frequency of 500 kHz?

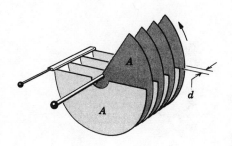

$A \equiv$ *area of plate*

$d \equiv$ *distance between plates*

Maximum capacitance is

$$C = \frac{(n-1)\,\epsilon_0 A}{d}$$

where n is number of plates.

Figure 15.10 A variable air capacitor.

Solution: Maximum current flows when

$$X_L = X_c$$

or

$$f = \frac{1}{2\pi\sqrt{LC}}$$

Solving for C we get

$$C = \frac{1}{4\pi^2 Lf^2}$$

$$= \frac{1}{4\pi^2(0.1\times10^{-3}H)\,(500{,}000\ s^{-1})^2}$$

$$= 1.01\times10^{-9}\,F$$

Power in AC Currents

We have seen in Section 15.4 that an LC circuit without resistance will oscillate forever, as energy is transferred back and forth between the inductor and capacitor. In this situation energy is stored either in the electric field of the capacitor, or the magnetic field of the inductor, and is never lost. Energy is only dissipated in a resistor. So the average power dissipated in the circuit of Figure 15.9 is

$$P_{av} = \frac{i_m^2}{2}R \tag{15.9}$$

The 2 comes from averaging over time.

Don't write

$$P_{av} = i_m^2 Z/2$$

Only R dissipates power. L and C do not enter the picture.

16 GEOMETRICAL OPTICS

16.1   Introduction

There are two types of optical phenomena studied in an introductory Physics course; they fall into categories called Geometrical Optics and Physical Optics. In geometrical optics the wavelength of the light is much smaller than the objects being observed. The light appears to move in straight lines, and produces sharp shadows of the objects. In physical optics one deals with much smaller objects. They are comparable in size to the wavelength of light.

This chapter deals only with geometrical optics. We will discuss

index of refraction

reflection

refraction

mirrors

lenses

16.2   Reflection and Refraction—Index of Refraction

The velocity of light in vacuum is 3×10^8 m/s. It becomes smaller in any material substance, its value given by the relation

$$c' = \frac{c}{n} \tag{16.1}$$

where n is called the index of refraction of the substance. n has different values for different substances -- look in your text for a chart.

Usually when light passes from one substance to another, two phenomena occur at the interface -- part of the light bounces back (is <u>reflected</u>) and part passes into the second medium, but is bent (is <u>refracted</u>). Figure 16.1 shows light moving from medium a into medium b.

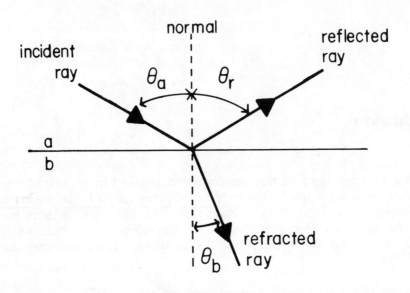

Figure 16.1

First observe the reflected wave in Figure 16.1. θ_a is called the angle of incidence and θ_r is called the angle of reflection.

$$\theta_a = \theta_r \qquad (16.2)$$

Now observe the refracted wave. θ_b is the angle of refraction. θ_a and θ_b are related by Snell's Law which states

$$n_a \sin\theta_a = n_b \sin\theta_b \qquad (16.3)$$

The above laws may be obtained experimentally. However, it is a remarkable fact that they can also be derived from the laws of electricity and magnetism.

* * * * * * * * * * * * * * *

Example 16-1

A coin lies at the bottom of a shallow lake; the depth of the water 70 cm. What is the apparent depth of the coin to the observer shown in Figure 16.2?

Assume that all rays entering the eye have angles of incidence and angles of refraction that are small (less than 10 degrees).

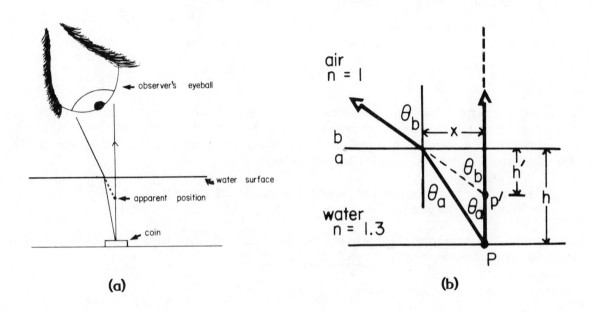

(a)   (b)

Figure 16.2

Solution: There is an extremely important principle involved here. To locate an image being formed by a surface or a lens system, draw at least two different rays from the object, and note where they meet, or <u>appear</u> to meet if extended backward. The image is at this point.

The first step in this problem is to apply Snell's Law to a typical ray, shown in Figure 16.2(b). Because the angles are small, we may replace sines by tangents.

$$(1.33) \sin \theta_a = (1.0) \sin \theta_b \quad \text{(Snell's Law)}$$

$$(1.33) \tan \theta_a = (1.0) \tan \theta_b \quad \text{(because angles are small)}$$

Applying trigonometry to Figure 16.2(b) gives

$$\tan \theta_a = \frac{x}{h}$$

$$\tan \theta_b = \frac{x}{h'}$$

Substituting these values of the tangents back into Snell's Law, we get

$$(1.33)\frac{x}{h} = (1.0)\frac{x}{h'}$$

$$h' = \frac{1.0}{1.33}h = 0.75\,h$$

We have not specified the angle of incidence in the above work; we have only required that it be small. Under this condition, all rays coming from point P appear (to the eye) to be coming from point P′. So the coin appears 0.75h below the surface.

* * * * * * * * * * * * * *

When light passes from one medium into a second which has a smaller index of refraction, the refracted ray bends away from the normal to the surface. From Figure 16.3 we see that there is a maximum angle of incidence that "makes sense". At this critical angle, the refracted ray passes parallel to the surface.

The angle of incidence for which the angle of refraction is 90° is called the critical angle.

For larger angles of incidence the incident ray is reflected back into the incident medium. This reflection phenomenon is called total internal reflection,

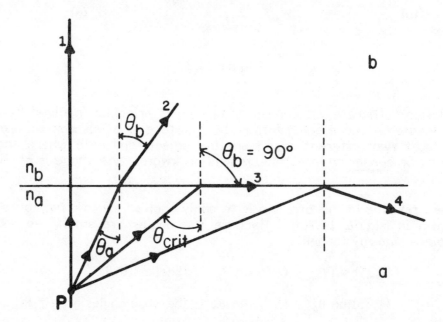

Figure 16.3

No other reflection process is as perfect as total internal reflection. For example, when light reflects from an ordinary mirror, some energy is lost in the mirror. It is total internal reflection that is the basis of modern fiber optics systems. By many such reflections light can be sent long distances without any appreciable loss of energy.

* * * * * * * * * * * * * *

Example 16-2

A point light source is 20 cm below a water-air surface. Find the radius of the largest circle at the surface through which light can emerge from the surface.

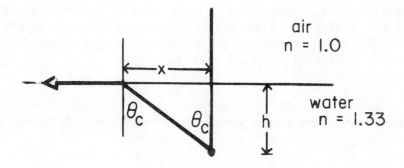

Figure 16.4

Solution: The first step is to get the critical angle. Then simple trigonometry (see Figure 16.4) gives the desired radius.

$$(1.33) \sin \theta_c = (1.0) \sin 90°$$

$$\theta_c = \sin^{-1}(0.75) = 48.6°$$

$$x = h \tan \theta_c$$

$$= (20 \text{ cm}) \tan (48.6°)$$

$$= 22.7 \text{ cm}$$

* * * * * * * * * * * * * *

16.3 Lenses and Mirrors

Lenses and mirrors is a terrible subject to cover in a book such as ours. The reason is that different textbooks and different instructors use slightly different sign conventions and notations. For example, some people write

$$\frac{1}{s} + \frac{1}{s'} = \frac{1}{f}$$

s	≡ object distance	
s'	≡ image distance	(16.4a)
f	≡ focal length	

for the thin lens formula, while others prefer

$$\frac{1}{p} + \frac{1}{q} = \frac{1}{f}$$

p	≡ object distance	
q	≡ image distance	(16.4b)
f	≡ focal length	

Sign conventions cause even more trouble. Enough little differences can cause total confusion in the mind of the reader. So in this section we have decided to follow a novel approach. We shall not do examples for you. Rather we shall present examples of lens problems from the examinations of several representative universities in the US. We thus give you a realistic idea of what to expect on a test. Try the problems. If you have difficulties, bug your professor. After all you're only spending about $12 for this book, but somewhere near $50,000 for tuition, room and board. Let your professor do some work! (The answers are given after all the problems are stated.)

Questions #1 and #2 are from a final exam for pre-meds at a large state university in the midwest.

1. An object is placed at a distance of 10 cm to the left of a diverging lens whose focal length is -5 cm. The image produced by the lens is

    (a)   real, erect and reduced.

    (b)   virtual, erect, and enlarged.

    (c)   real, inverted, and enlarged.

    (d)   virtual, erect, and reduced.

    (e)   virtual, inverted, and reduced.

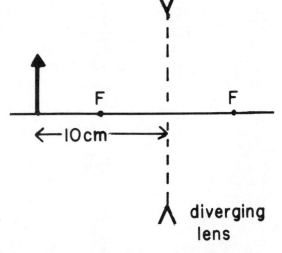

2. A microscope can be assembled from two identical, positive lenses each having a focal length of f and separated by a distance d. The lateral magnification attainable with the arrangement

    (a)   is necessarily equal to 1.0.

    (b)   can never be greater than d/f.

    (c)   depends only on the value of f.

    (d)   has no upper limit

    (e)   depends only on the object distance.

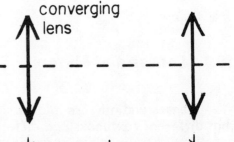

Question #3 is from a final exam for engineers at an Ivy League university.

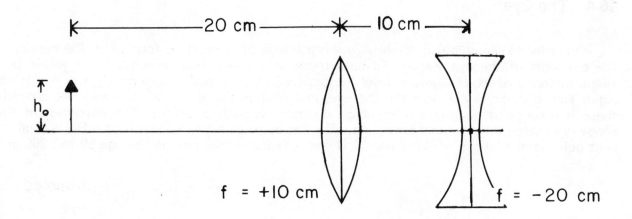

3.   Two thin lenses and an object are arranged as shown above.

   (a)   Where (relative to the position of the converging lens) is the "image" produced by the converging lens?

   (b)   What is the lateral magnification of the converging lens in this situation? Is the image real or virtual?

   (c)   Where (relative to the position of the diverging lens) is the final image produced by both lenses?

   (d)   What is the total lateral magnification of the system of lenses? Is the final image real or virtual? Inverted or upright relative to the initial object?

Question #4 is from a final exam for pre-meds at a large university in the south.

4.   How far from a 50 mm focal length lens must an object be placed if its image is to be magnified 3.0x and be real?

   (a) 6.7 cm   (b) 5.0 cm   (c) 3.3 cm   (d) 4.66 cm   (e) 0.04 cm

Answers

1:   (d)   3(a):   20 cm to right of converging lens   4:   (a)

2:   (d)   3(b):   lateral magnification has magnitude 1; image is real.

                  3(c):   20 cm to right of diverging lens

                  3(d):   lateral magnification has magnitude 2; final image is real

16.4   The Eye

In order to see an object clearly, a sharp image of it must be formed on the retina. If the elements of the eye were rigid (like those of a cheap box camera), there would be a single distance at which sharp retinal images could be formed. Fortunately the eye is not rigid; strong muscles can squeeze the eye, and change the shape of the lens. As a result, there is a range of distances over which distinct vision is possible. The extremes of this range are called the near and far points. The far point of a normal eye is at infinity, and the near point is at ~10 cm, while at age 50 it has receded to ~40 cm, and by age 60 to ~200 cm!

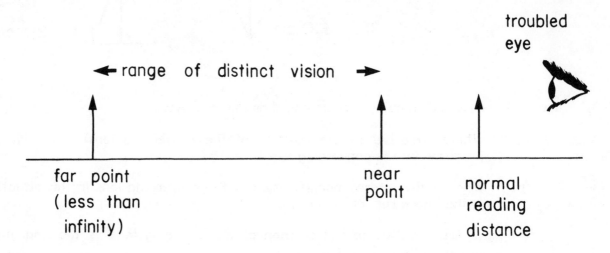

Figure 16.5

In Figure 16.5 we show a schematic diagram of a troubled eye. Glasses can correct the fact that the far point is less than infinity. A diverging lens can "bring in" objects from infinity to the far point so that they become visible (see Example 16-3).

Similarly, the fact that the near point is farther than the normal reading distance can be helped --

1.   Hold the reading material farther from the eye. This approach has its limitations, since small print will become difficult to resolve.

2.   Wear glasses with the correct converging lens. These will "move" reading material out to the near point so that it becomes visible, and sufficiently large (see Example 16-4).

At this point you must understand that a single lens cannot be both converging (to correct near point difficulties) and diverging (to correct far point difficulties). When your near point distance exceeds the lengths of your arms, YOU ARE IN TROUBLE! A single lens won't help you. It is time for BIFOCALS. Remember that the near point of the eye recedes with age, so the requirement for bifocals is a normal consequence of living more than 40 or 50 years. (It is not to be confused with senility.)

Example 16-3

The far point of a certain eye is 1.5 m in front of the eye. What eyeglass lens will allow this eye to see an object at infinity clearly?

Solution: The object is at infinity. We wish the image to be at the far point. Using (more or less) standard notation, we have

$$\text{object distance} = \infty$$

$$\text{image distance} = -1.5 \text{ m}$$

$$\frac{1}{\infty} + \frac{1}{-1.5 \text{ m}} = \frac{1}{f}$$

Answer: A diverging lens with focal length 1.5 m.

* * * * * * * * * * * * * *

Example 16-4

The near point of a certain eye is 100 cm in front of the eye. What eyeglass lens should be used in order to see clearly an object 25 cm in front of the eye?

Solution: In this case

$$\text{object distance} = +25 \text{ cm}$$

$$\text{image distance} = -100 \text{ cm}$$

$$\frac{1}{+25 \text{ cm}} + \frac{1}{-100 \text{ cm}} = \frac{1}{f}$$

Answer: A converging lens with focal length 33 cm.

17 PHYSICAL OPTICS

17.1 Introduction

In physical optics you are observing objects which are comparable in size to the wavelengths of light. Instead of treating light as rays, you must treat it as a true <u>wave</u>. The principles of wave motion as discussed in Chapter 10 now hold sway.

In geometrical optics you had straight streaks of light shooting through and bouncing off objects, with well defined angles of incidence, reflection, and refraction. In physical optics, light is treated as a wave and things become less clear-cut. Light bubbles through cracks and curls around obstacles. Few people realize that the strange phenomena of physical optics have important applications in science and technology. Interference and diffraction are just about the only ways that we have to measure very small distances, ranging from cracks in airplane wings to the distances between atoms in DNA.

17.2 Interference of Light

In Section 10.5 we studied the case of two speakers emitting sound in phase (coherently). One obtained maxima and minima at different points in space, depending on the path differences. The relevant equation was (10.11)

$$\Delta\phi = \frac{2\pi}{\lambda}\Delta x \qquad\qquad (10.11)$$

where $\Delta\phi$ was the phase difference and Δx was the path difference. Light behaves the same way with one important difference -- you cannot create two coherent light sources. In other words, you cannot make two light bulbs emit light in phase, because the emission of light is a random process. Electrons in a light bulb are frantically hopping about out of our control.

The way in which you get two coherent light sources is to start with a single source and artificially break it up into two sources. We show three ways that this can be done in Figure 17.1.

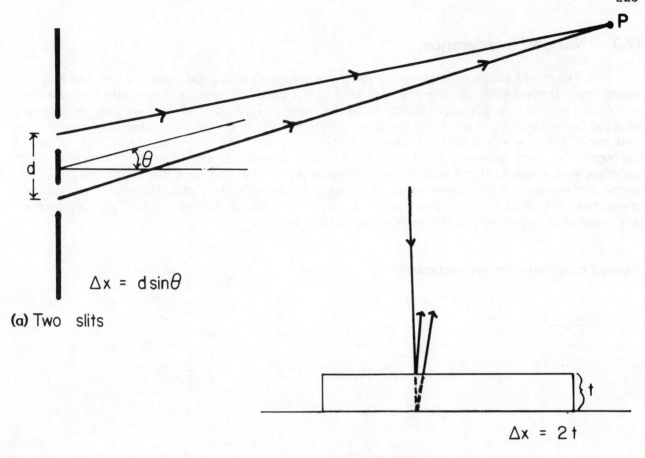

$$\Delta x = d\sin\theta$$

(a) Two slits

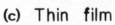

$$\Delta x = 2t$$

(c) Thin film

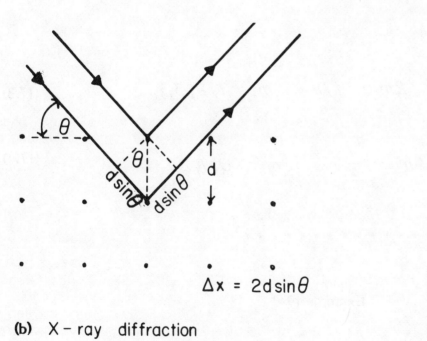

$$\Delta x = 2d\sin\theta$$

(b) X-ray diffraction

Figure 17.1

17.3 "Normal" Interference

In Figure 17.1(a) we have artificially created two coherent sources, by sending the wave front from a single source through two slits. The slits behave coherently because they are really part of a single wave front. In both Figures 17.1(b) and 17.1(c) we create coherent sources by reflecting from two (or more) surfaces. Each surface behaves like a source of radiation, and there is a definite phase relation between the emissions from the different surfaces. So they behave coherently. In all three situations, when the waves recombine at an observation point, there will be an intensity pattern of maxima and minima. It is the phase difference $\Delta\phi$ that determines what we see, but it is the path difference Δx that one measures. We shall refer to those cases in which the phase difference and path difference are related by Equation (10.10), as "normal interference".

Normal conditions for interference --

Maxima:

$$\Delta\phi = 2m\pi \qquad m = 0, 1, 2, \ldots \qquad (17.1)$$

$$\Delta x = m\lambda \qquad m = 0, 1, 2, \ldots \qquad (17.2)$$

Minima:

$$\Delta\phi = (2m - 1)\pi \qquad m = 1, 2, 3, \ldots \qquad (17.3)$$

$$\Delta x = (m + \frac{1}{2})\lambda \qquad m = 0, 1, 2, \ldots \qquad (17.4)$$

* * * * * * * * * * * * * *

Example 17-1

Light of wavelength 400 nm passes through a double slit configuration, as shown in Figure 17.1(a). The distance between the slits is 700 nm. At what angles θ will one observe the first and second order maxima?

Solution: Using Figure 17.1(a) and Equation (17.2), we see that the conditions for maxima are

$$d \sin \theta = m\lambda$$

$$(700) \sin \theta = m (400)$$

To get the first and second order maxima, we set m = 1.2.

m = 1 $\qquad \qquad \sin \theta = \dfrac{400}{700}; \quad \theta = 34.8°$

m = 2 $\qquad \qquad \sin \theta = \dfrac{800}{700}; \quad \theta = $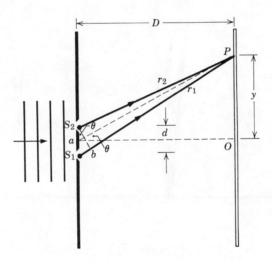

There is an important lesson here. If λ becomes significantly larger than d, there are no maxima or minima because sin θ is greater than unity (1.0) in all cases. We no longer get a unique interference pattern, and the width of the slit is not measurable. <u>To measure small distances, you need even smaller wavelengths.</u> This is the reason that X-ray diffraction is used to determine interatomic distances; X-rays are easily obtained radiation with wavelengths smaller than these distances.

One more point concerning double slit interference. Often problems refer to a special setup called Young's experiment (see Figure 17.2).

Figure 17.2 Rays from S_1 and S_2 combine at P. Actually, D >> d, the figure being distorted for clarity. Point a is the midpoint of the slits.

In this case the angle θ is always small (θ is less than ~10°), and to very high accuracy the following formula holds:

$$\frac{\lambda}{d} = \frac{y}{D} \qquad (17.5)$$

A major error made by students is using this formula when it is not valid. The example you have just studied is such a case!

* * * * * * * * * * * * * *

17.4 "Abnormal" Interference

The fact that different substances can have different indices of refraction complicates the conditions for interference. The wavelength of light in a material with index of refraction n is

$$\lambda' = \frac{\lambda}{n}$$

where λ is the wavelength in vacuum. If this were the only change, it would be barely worth mentioning. However, a second phenomena occurs. The laws of electricity and magnetism tell us that when light reflects off the interface between substance a and substance b (as shown in Figure 17.3), there is a phase change of π radians introduced if $n_b > n_a$.

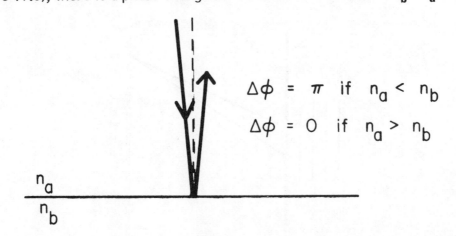

$$\Delta\phi = \pi \quad \text{if} \quad n_a < n_b$$
$$\Delta\phi = 0 \quad \text{if} \quad n_a > n_b$$

n_a
n_b

Figure 17.3

This is a serious new complication. For example, the phase change in the process described by Figure 17.1(c), no longer depends on just the <u>path</u> difference. In fact, the normal conditions for interference may be completely reversed. We refer to the conditions for maxima and minima in the case where an additional phase change of π radians is introduced by reflection, as abnormal conditions.

Abnormal conditions for interference --

Maxima:

$$\Delta\phi = 2m\pi \qquad m = 0, 1, 2, \ldots \tag{17.6}$$

$$\Delta x = (m + \frac{1}{2})\lambda \qquad m = 0, 1, 2, \ldots \tag{17.7}$$

Minima:

$$\Delta\phi = (2m - 1)\pi \qquad m = 1, 2, 3, \ldots \tag{17.8}$$

$$\Delta x = m\lambda \qquad m = 0, 1, 2, \ldots \tag{17.9}$$

* * * * * * * * * * * * * * *

Example 17-2

In Figure 17.4 we show several setups in which interference occurs. In each case, state the conditions for maxima and minima.

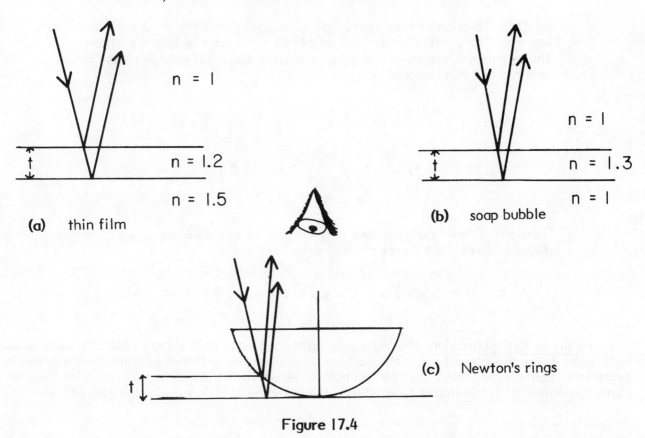

(a) thin film

(b) soap bubble

(c) Newton's rings

Figure 17.4

Solution: (a) This is a <u>normal</u> situation. A phase change of π radians occurs at <u>both</u> the top and bottom surfaces of the thin film, so no difference in phase is introduced by the reflection. But note that the extra path length occurs in the medium with index of refraction n = 1.2; so the conditions for maxima and minima are

Max: $\qquad 2t = \dfrac{m\lambda}{1.2} \qquad\qquad m = 0,1,2,...$

Min: $\qquad 2t = \dfrac{(m + \frac{1}{2})\lambda}{1.2} \qquad m = 0,1,2,...$

(b) This is an <u>abnormal</u> situation. A phase change of π radians occurs at the upper surface, but not at the lower surface. The extra path length occurs in the medium with index of refraction n = 1.3, so the conditions for maxima and minima are

Max: $\qquad 2t = \dfrac{(m + \frac{1}{2})\lambda}{1.3} \qquad m = 0,1,2,...$

Min: $\qquad 2t = m\dfrac{\lambda}{1.3} \qquad\qquad m = 0,1,2,...$

(c) This situation is also <u>abnormal</u>. No phase change occurs at the upper surface, but a phase change of π occurs at the lower surface. The extra path difference occurs in air this time. The conditions for maxima and minima are:

Max: $\qquad 2t = (m + \frac{1}{2})\lambda \qquad m = 0,1,2,...$

Min: $\qquad 2t = m\lambda \qquad\qquad m = 0,1,2,...$

Caution: There is a minimum as $t \to 0$. The eye looking down from above will see a dark spot at the center.

* * * * * * * * * * * * * *

In <u>single slit diffraction</u>, the light passing through one half (of a single slit) interferes with light from the other half. The full mathematical description of this phenomenon is relatively complicated, and usually is not covered in an introductory Physics course. Rather, <u>single slit diffraction</u> is treated as a special case of "abnormal interference".

18 MODERN PHYSICS

18.1 Introduction

If your instructor and textbook are typical, you have spent most of your Physics course covering roughly forty chapters of classical (pre-twentieth century) Physics. These chapters include the subjects of mechanics, heat and thermodynamics, wave motion, and electricity and magnetism. Now there are only three weeks remaining in the course to cover all of modern (twentieth century) Physics!

There is a problem. Modern Physics introduces many radical new concepts, such as the uncertainty principle, wave particle duality, quantization, and special relativity. Each of these ideas changed the course of scientific thought, and most earned their author a Nobel prize. The main lesson that the average person may glean from these ideas, is that the behavior of very fast objects (moving near the speed of light), or very small objects (electrons, protons, etc.), cannot be understood on the basis of classical Physics. This is because, in many cases, the phenomena of modern Physics require a "picture" of nature that must be inferred from mathematical equations. It took the world's scientific community roughly one half century to absorb these ideas. The problem we referred to earlier is that you will have to learn them in three weeks!

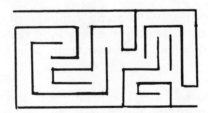

Our purpose in this chapter is to guide you through the maze of information presented in the modern Physics sections of most introductory texts. We shall isolate and develop the key ideas, and discuss problems that are likely to appear on exams. The good news is that the complexity of the subject offers certain unexpected advantages. For example, in order to be at all soluble, problems must be more straightforward than in earlier parts of the course. The bad news is that in a short survey of modern Physics, you will never feel secure with the subject material - you will always have that queasy feeling that you are only seeing the tip of the iceberg. But you can easily learn to live with this feeling. After all, you have already gone through courses such as biology and chemistry, where most of the fundamental ideas were beyond your grasp at the time.

We shall begin by discussing those aspects of modern Physics which require concepts beyond those of classical Physics for their development. We shall also consider the corresponding classical principles which fall by the wayside. The four major topics covered in this chapter are

Special Relativity

Wave-Particle Duality

Quantization

Uncertainty Principle

It is for better understanding that we cover these four topics in the order indicated. The most likely problems to appear on examinations will involve atomic spectra (quantization) and special relativity (if you get to these subjects in lecture).

18.2   Special Relativity

The velocity of electromagnetic waves in a vacuum ($c = 3.0 \times 10^8$ m/s) is the same for all observers. It is the maximum velocity with which information can be transferred from one point in space to another. (Remember that light is an electromagnetic wave.)

A law of nature ....

Unless you are already familiar with some aspects of Einstein's Special Theory of Relativity, you should be deeply disturbed by the previous statement. It completely violates your previous experience with velocities, derived from seventeen or more years of being alive.

Were you disturbed? Of course not! The problem is that you may have had courses in art appreciation and music appreciation, but never Physics appreciation. You have never learned to appreciate the world around you and cannot be disturbed by something you do not appreciate (such as the behavior of velocities). Unfortunately one cannot understand and apply the concepts of Modern Physics without special "appreciation training" which we shall (fortunately) provide in this chapter.

Lets start with the idea that a velocity can appear the same to all observers. On the face of it, such a statement is patently absurd. The velocity or speed of an object is a relative quantity. Its value is different for different observers, depending on their motion relative to the object. For example, suppose that you are sitting in a train next to a friend. The train is traveling at 60 mph, and passes a person standing alongside the track. That person would say that your friend had a speed of 60 mph, while you of course would say that your friend had zero speed (relative to you).

Suppose your friend began walking on the train. Can you imagine a world in which the person by the track and you, would <u>both</u> perceive your friend to be moving at the same speed? The remarkable fact is that there is such a world -- it is our world for speeds that near that of light. As you might guess, such behavior has profound consequences for the laws of Physics; these were first proposed by Albert Einstein in his Special Theory of Relativity in 1905. In order for light to behave as it does, Newton's laws must begin breaking down at some point. For example velocities cannot add in what we consider a normal way under all circumstances.

Einstein Addition Law for Velocities ....

The velocities we experience on earth are very small compared to that of light, so you should not be surprised that at near-light velocities, phenomena occur that violate our experience. In order for the velocity of light to be constant for all observers, the very nature of space and time must be different from what we observe everyday. In Figure 18.1 we show a rocket ship moving at velocity V with respect to a stationary observer, and on the rocket ship a person runs with velocity v'.

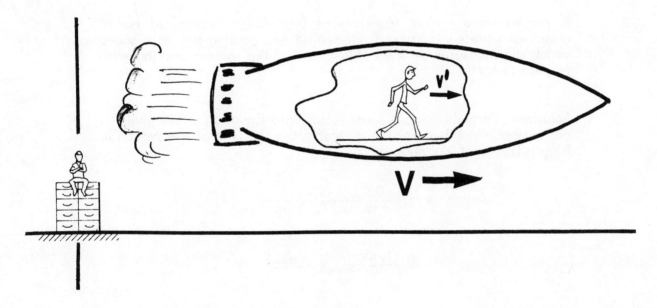

Figure 18.1

According to the Physics of Newton (low velocities) the velocity of the person on the rocket is

$$v = V + v' \qquad (18.1)$$

relative to the stationary observer. However according to special relativity, the <u>addition</u> <u>law</u> <u>for</u> <u>velocities</u> is

$$v = \frac{V + v'}{1 + Vv'/c^2} \qquad (18.2)$$

Equation (18.1) implies that there is no maximum velocity, because we can always "boost" to a higher velocity by having a person start walking on the object that is assumed to have the maximum velocity. On the other hand, Equation (18.2) guarantees that v will always be less than c if V and v' are each less than c. More amazing is the fact that v is equal to c if either V or v' is equal to c. [Substitute c for v' in Equation (18.2) and see what happens.] When both V and v' are small compared to c, we can neglect Vv'/c compared to unity, and we recover Eq. (18.1) from Eq. (18.2).

We have not proven anything in the above discussion. We merely wish you to appreciate the fact that there are kinematics equations which allow the velocity of light to be constant for all observers, and which still satisfy Newton's laws at everyday velocities.

* * * * * * * * * * * * * *

Example 18-1

A person on a rocket traveling at 0.6c (with respect to earth) observes a meteor coming from behind and passing her at a speed which she measures as 0.8c. How fast is the meteor moving with respect to earth?

Solution: Straightforward substitution into Equation (18.2) gives us our answer. In this case, the meteor plays the role of the person in Figure 18.1 -- it has velocity v'.

$$v = \frac{0.60c + 0.80c}{1 + (0.60c)(0.80c)/c}$$

$$v = \frac{1.40c}{1 + 0.48} = 0.95c$$

* * * * * * * * * * * * * *

The addition laws for velocities is not all that breaks down at near-light velocities. In Figures 18.2 and 18.3 we show direct pictorial evidence that the Newtonian formulation of energy conservation and momentum conservation no longer holds. The tracks in Figure 18.2 describe the elastic collision of a moving proton with a second proton, initially at rest. In this case v = 0.1c and we expect Newton's laws to be fairly accurate. Indeed the angle between the two outgoing protons is 90 degrees, as explained in Example 6-6. In Figure 18.3 we have an elastic collision of a moving electron with a second electron at rest. In this case v = 0.97c, and you can see that the outgoing electrons no longer obey Newton's laws. The angle between the two electrons is much less than 90 degrees, and (you must take our word) is given precisely by the theory of special relativity.

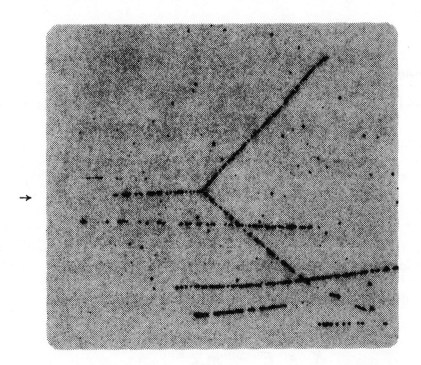

Figure 18.2 Slow protons collide like billiard balls.

Figure 18.3 The curved path is due to the presence of a magnetic field used to determine the momentum of each electron.

Mass-energy equivalence ....

    The most famous consequence of the special theory of relativity is the mass-energy relation.

$$E = \frac{m_o c^2}{\sqrt{1 - v^2/c^2}}$$ *(m_0 is the old fashioned mass you use everyday)*   (18.3)

If you get a special relativity question on an exam, it will most likely involve this relation. According to Equation (18.3) an object of mass m_0 has energy $m_0 c^2$ even when the velocity is zero. Hence $m_0 c^2$ is called the "rest mass energy". The kinetic energy of a "high speed" particle is defined to be the total energy minus the rest mass energy --

$$K = E - m_o c^2$$ (18.4)

$$\approx \frac{1}{2} m_o v^2 \quad (for \quad v << c)$$ (18.5)

We give you the relation Equation (18.5) to show you that at everyday velocities, one gets the familiar expression for the kinetic energy. [Eq. (18.5) can be derived from Eq. (18.4) using either the binomial expansion for students without sufficient calculus, or the Taylor expansion for students with enough calculus.] Examination questions are often designed to punish you for using ½ mv² for the kinetic energy at velocities near that of light.

* * * * * * * * * * * * * * *

Example 18-2

An electron initially at rest is accelerated through a difference of potential of 80,000 volts. What is the final velocity of the electron? ($m_0 c^2$ = 500,000 eV for an electron.)

Solution: You have already been warned in Example 12-7 to watch out for special relativity. Using conservation of energy, we set the loss in potential energy equal to the gain in kinetic energy, and solve for v.

    A major simplification occurs in this and many Modern Physics problems that you will encounter, <u>if you work with energies in units of electron volts</u>.

$$\frac{m_0 c^2}{\sqrt{1 - v^2/c^2}} - m_0 c^2 = e\,\Delta V$$

$$\frac{500,000 \text{ eV}}{\sqrt{1 - v^2/c^2}} - 500,000 \text{ eV} = 80,000 \text{ eV}$$

$$\frac{5}{\sqrt{1 - v^2/c^2}} = 5.8 \quad \longleftarrow \quad \text{(square both sides and solve for } v)$$

$$v = 0.507c = 1.52 \times 10^8 \text{ m/s}$$

So you see that our answer in Example 12-7 was off by 10%.

* * * * * * * * * * * * * *

The other type of "$E = mc^2$" problem that often appears on examinations involves the concept of mass defect. In a nuclear or elementary particle reaction, the mass of the final particles is often less than that of the initial particles, and thus energy is released according to $E = \Delta mc^2$. Δm is the difference between the masses in the initial and final states, and is called the mass defect. Such problems are covered thoroughly in your text, and we shall not give examples here. However we wish to warn you about one common error. The mass defect is usually calculated from tables which give nuclei masses in atomic mass units. You must convert atomic mass units to kilograms for use in $E = \Delta mc^2$. Or, you can use the fact that if the mass is 1 atomic mass unit, mc^2 is 931 meV. Effectively

$$1 \text{ atomic mass unit} \equiv 931 \text{ meV}$$

18.3   Wave Particle Duality

In Classical Physics, waves and particles are distinctively different.

Particles are precisely localized structures, and their collisions are described by Newton's laws.

On the other hand the energy in a wave is not precisely localized, and "collisions" of waves are described by the superposition principle for the wave amplitudes. In addition, waves exhibit the phenomena of interference and diffraction (particles do not). Transverse waves can be polarized (particles cannot).

Special relativity tells us that there is a relation between mass and energy. With such a relation, the distinction between particles and waves become unclear. For example, in Figure 18.4(a) we plot the mass density ρ of a small object of width a, as a function of position. As we might guess, the mass density is zero everywhere except in the small region of space occupied by the object.

On the other hand, consider a sinusoidal electromagnetic wave moving through space in the x direction. In Figure 18.4(b) we plot the energy density U divided by c^2 as a function of x (at a particular time). According to Einstein, an energy density U divided by c² is equivalent to a mass density ρ. Still, the sinusoidal wave in Figure 18.4(b) does not behave like a particle, because the equivalent mass is spread out over all space. In other words Figure 18.4(a) and 18.4(b) look nothing like each other.

Finally, in Figure 18.4(c) we plot the equivalent (energy density)/c² in an electromagnetic wave pulse, which can be constructed by adding many sinusoidal waves of adjacent wavelengths. In this case the electrical energy, and thus the equivalent mass, is concentrated in a small region of space. Such a localized concentration of energy does have features of a particle. Research in this area was associated with several Nobel prizes - Einstein, de Broglie, Planck.

The Photoelectric Effect

In 1921 Albert Einstein received the Nobel prize in Physics for his services to Theoretical Physics, and especially for his discovery of the law of the photoelectric effect.

In the photoelectric effect, electrons are ejected from the surface of a metal by incident light. Only light with "sufficiently high" frequency will do the job. If the frequency is "too small," no matter how intense the light, no photoelectrons will appear.

In 1905 Einstein explained the photoelectric effect with the remarkable assumption that, under some circumstances, light behaves as if its energy is concentrated into particles called photons. The energy of a single photon is given by

$$E = h\nu$$
$$\nu = \frac{c}{\lambda} \tag{18.6}$$

where ν is the frequency of the light, and h is Planck's constant. Single photons (never fractions of a photon) are absorbed or emitted by charged particles. Applying Equation (18.6) to the photoelectric effect, Einstein obtained the relation

$$K_{max} = h\nu - \phi \tag{18.7}$$

where $h\nu$ is the energy of the photon striking the metal, and ϕ is called the work function of the metal. K_{max} represents the maximum kinetic energy that the electron can have outside the metal, and the work function ϕ represents the minimum work needed to overcome the forces binding the electron in the metal.

* * * * * * * * * * * * * * *

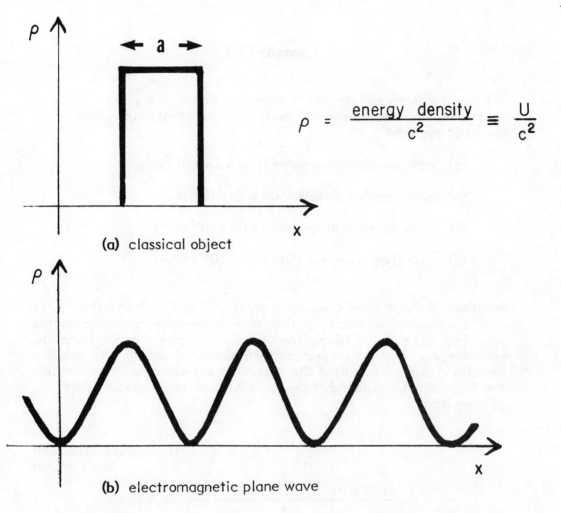

$$\rho = \frac{\text{energy density}}{c^2} \equiv \frac{U}{c^2}$$

(a) classical object

(b) electromagnetic plane wave

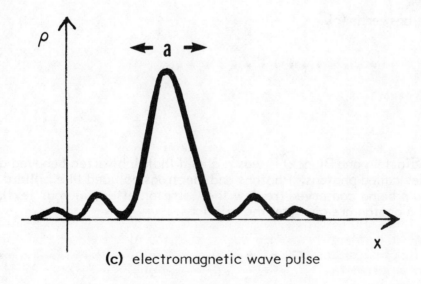

(c) electromagnetic wave pulse

Figure 18.4 (c) is like (a) if you cannot observe wiggles.

Example 18-3

A clean tungsten surface has a work function ϕ of 4.52 eV. We wish to obtain photoelectrons with kinetic energies greater than 1.60 eV. We must use light

(a)   with wavelength greater than 4.28×10^{-7} m

(b)   with wavelength less than 4.28×10^{-7} m

(c)   with wavelength less than 2.04×10^{-7} m

(d)   with frequency less than 0.70×10^{15} Hz

Solution: We wish kinetic energies greater than a certain value. This means we want frequencies greater than some value, and wavelengths less than some value (since frequency is inversely proportional to wavelength). Thus we immediately eliminate (a) and (d). We now use Equation (18.7) to calculate the maximum wavelength. You simplify the calculation considerably by using h in units of eV-s. Since $\nu = c/\lambda$, we have

$$\frac{hc}{\lambda} = K_{max} + \phi \qquad (K = 0 \text{ for maximum wavelength})$$

$$\frac{(4.14 \times 10^{-15} \text{eV-s}) (3 \times 10^{8} \text{ m/s})}{\lambda} = 6.12 \text{ eV}$$

$$\lambda = 2.04 \times 10^{-7} \text{ m}$$

So the correct answer is (c).

* * * * * * * * * * * * * *

Matter Waves

By 1924 (thanks to Einstein and Planck) it was realized that light often behaved as if it were composed of particles called photons. Photons and electrons collided like billiard balls, with energy and momentum being conserved (review the Compton effect in your text). The energy and momentum of a photon are given by the relations

$$E = h\nu \tag{18.6}$$

$$p = \frac{h}{\lambda} \tag{18.8}$$

In 1924 Louis de Broglie of France (Nobel prize, 1929) reasoned that there should be a symmetry in nature. If waves could behave like particles under certain circumstances, then particles should behave like waves under certain circumstances. From the energy-momentum relations for photons, de Broglie decided that the frequency of a particle wave should be given by

$$\nu = \frac{E}{h}$$

(18.9)

and the wavelength by

$$\lambda = \frac{h}{p}$$

(18.10)

De Broglie's hypothesis may be checked experimentally. Electrons with an energy of 100 eV have a de Broglie wave length of 12 nm, which is roughly the size of the spacing in atomic crystals. If one "shines" such electrons on a crystal, one should get a diffraction pattern identical to the one obtained with X-rays of the same wavelength. This prediction was confirmed.

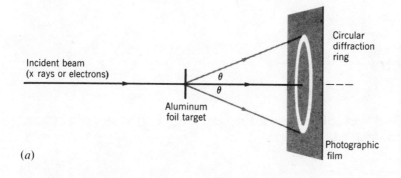

(a)

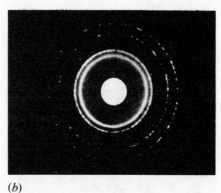

(b)

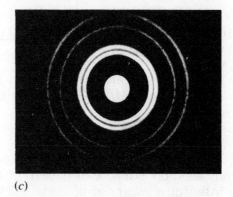

(c)

Figure 18.5 (a) an experimental arrangement for producing a diffraction pattern characteristic of an aluminum target. (b) Pattern for an incident x-ray beam. (c) Pattern for an incident electron beam. The electron energy has been chosen so that the de Broglie wavelength of the electron beam is approximately equal to the wavelength of the x-rays employed. The central spot, identifying the directly transmitted beam, has been deleted in each case, (b and c) From the PSSC film Matter Waves, Production #68 Education Development Center. Newton, MA.

Example 18-4

Find the de Broglie wavelength for a beam of electrons whose kinetic energy is 100 eV.

Solution: The energy of these electrons is so small that we do not have to worry about special relativity. So we may write

$$\lambda = \frac{h}{p}$$

$$= \frac{h}{\sqrt{2mE}} \qquad \left(\text{since } E = \tfrac{1}{2}mv^2 = \frac{p^2}{2m}\right)$$

$$\lambda = \frac{6.6 \times 10^{-34} \text{J-s}}{\sqrt{2(9.1 \times 10^{-31} \text{kg})(100 \text{ eV})(1.6 \times 10^{-19} \text{J/eV})}}$$

$$\lambda = 1.2 \times 10^{-10} \text{m}$$

* * * * * * * * * * * * * *

18.4 Quantization—Atomic Spectra

In Classical Physics most dynamical variables are continuous; they can take on a range of values which are as close together as desired.

We can walk at any speed (allowed by muscular strength).

A satelite can orbit at any radius. There are no regions of space forbidden to it.

On the other hand, we run into quantization when we consider waves on a string fixed at both ends.

The wavelengths that can propagate on a string fixed at both ends are quantized. In other words they can only take on special or discrete values.

Quantization shows up in the less traditional phenomena of Modern Physics. We have already seen an example of quantization in the photoelectric effect. Electrons could absorb only integral numbers of photons. Quantization once again shows up when one examines the spectrum of light emitted by atoms. The fact that line spectra appear is well known, but cannot be explained in the boundaries of Classical Physics. Not only does the classical theory of electricity and magnetism disallow the possibility of line spectra; it also precludes the traditional picture of an atom in which electrons circle the nucleus. An electron moving around an atom has a centripetal acceleration, and according to the laws of electricity and magnetisim, a charged particle radiates energy when it accelerates. Thus the electrons should lose energy and fall into the nucleus in a very short time. Of course we know that this does not happen!

If we use our knowledge that electrons can behave like waves with a de Broglie wavelength, a possible explanation of atomic structure begins to emerge. Suppose we have a string fixed at both ends, and we try to get transverse waves of a given wavelength to propagate on the string. Most wavelengths will die out almost at once. They bounce back and forth on the string, and eventually destroy themselves through interference. The only ones that remain forever, are those for which standing waves can be created. Let's assume that a similar picture holds for electrons in orbit about a nucleus. The only de Broglie wavelengths that can exist forever, are those that can fit into an orbit (see Figure 18.6).

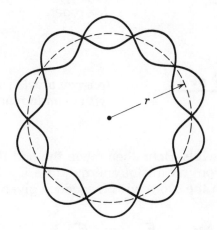

Figure 18.6 Showing how an electron wave can be adjusted in wavelength to fit an integral number of times around the circumference of a Bohr orbit of radius r.

If the wave did not fit into the orbit as shown in Figure 18.6, there would be a mismatch, and the circulating wave would eventually destroy itself by interference. In other words, the wavelength must precisely fit around the circumference of the orbit an integral number of times:

$$n\lambda = 2\pi r \quad n = 1,2,\ldots \tag{18.11}$$

Using the de Broglie relation [Eq. (18.10)], one can express the wavelength in terms of the momentum

$$pr = \frac{nh}{2\pi} \qquad n = 1, 2, \ldots \qquad (18.12)$$

where pr is the angular momentum of the electron about the proton. Equation (18.12) is the Bohr quantization condition. It was derived by Niels Bohr long before the concept of matter waves was developed by de Broglie. How did Bohr do it? The answer is that he was able to do it because he was Niels Bohr -- a genius!

We now obtain the Bohr semi-classical model of the hydrogen atom by combining the non-classical Bohr quantization condition with simple results from classical electrostatics. For an electron in a circular orbit about a proton, Newton's Second Law plus Coulomb's Law, plus the Bohr quantization condition gives (see your textbook for detailed derivation):

$$E_n = \frac{-13.6 \ eV}{n^2} \qquad \text{(energy levels for hydrogen)} \qquad (18.13)$$

$$E_n = \frac{-Z^2(13.6 \ eV)}{n^2} \qquad \text{(energy levels for nucleus with atomic number Z)} \qquad (18.14)$$

Having obtained these energy levels, Bohr then made the additional assumption that when an electron makes a transition from an initial energy state E_1 to a lower energy state E_2, a single photon was emitted from the atom with frequency given by

$$h\nu = E_2 - E_1$$
$$= -Z^2(13.6 \ eV)\left[\frac{1}{n_2^2} - \frac{1}{n_1^2}\right] \qquad (18.15)$$

The same frequency relation holds when an electron makes a transition from a lower energy level to a higher energy level, except in this case the electron absorbs a photon.

The Bohr model predicts many atomic spectra with surprising accuracy. It was marvelous in its time, just as the first airplane built by the Wright brothers was marvelous in its time. But the Bohr model is not really a theory based on fundamental principles; rather it is a hodge-podge of classical and modern ideas held together loosely by genius.

The fundamental theory came a bit later. Today we know that classical mechanics and classical electricity and magnetism break down in the atomic realm. A new theory of atomic phenomena called quantum mechanics was developed in the 1920's. In quantum mechanics atoms and molecules are described by a wave equation (called the Schroedinger Equation). Ideas like the wave particle duality, and the Bohr quantization condition emerge naturally from this theory.

Example 18-5

The He$^+$ ion is a helium nucleus with a single orbiting electron. A helium nucleus (sometimes called an α-particle) consists of two protons and two neutrons.

(a) Construct an energy level diagram showing the lowest four levels.

(b) What is the shortest wavelength photon which can be absorbed by the He$^+$ when it is in its ground state, and still have the electron bound to the nucleus?

(c) If an electron in the lowest energy level absorbs a photon with wavelength 1,000 nm, with what kinetic energy will it emerge from the ion?

Solution: (a) The He$^+$ ion has atomic number Z = 2; the lowest four levels are given by Equation (18.14) with n = 1,2,3,4.

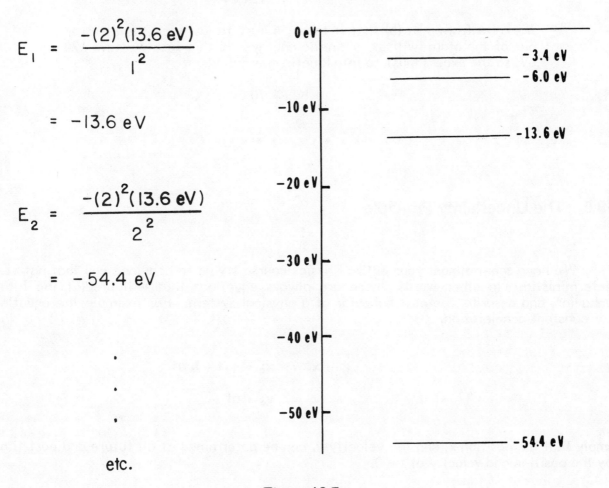

$$E_1 = \frac{-(2)^2(13.6\ eV)}{1^2}$$

$$= -13.6\ eV$$

$$E_2 = \frac{-(2)^2(13.6\ eV)}{2^2}$$

$$= -54.4\ eV$$

⋮

etc.

Figure 18.7

(b)   The largest energy that the electron can have and still be in an orbit is just under zero eV (see Figure 18.7). If the electron in its ground state absorbs a photon with 54.4 eV it will wind up with a total energy of zero eV. The corresponding wavelength is

$$\frac{hc}{\lambda} = 54.4 \text{ eV}$$

$$\lambda = \frac{(4.14 \times 10^{-15} \text{eV-s})(3 \times 10^{8} \text{ m/s})}{54.4 \text{ eV}}$$

$$\lambda = 0.228 \times 10^{-7} \text{ m}$$

(c)   Let's calculate the energy of a 10 nm photon.

$$E = \frac{hc}{\lambda} = \frac{(4.14 \times 10^{-15} \text{eV-s})(3 \times 10^{8} \text{m/s})}{10 \times 10^{-9} \text{m}}$$

$$E = 124 \text{ eV}$$

We know from part (b) that it takes 54.4 eV to just kick the electron out of the atom (with zero kinetic energy). In this case we have 124 eV, so the excess must go into kinetic energy.

$$K \approx 70 \text{ eV}$$

* * * * * * * * * * * * * * *

18.5   The Uncertainty Principle

We have spent almost your entire Physics course trying to convince you that nature is deterministic. In other words, there are always equations that will predict the future behavior, and describe the past behavior of a physical system. For example the equations for constant acceleration,

$$x = x_0 + v_0 t + \tfrac{1}{2} at^2$$

$$v = v_0 + at$$

imply that the position x, and the velocity v, can be determined at all future and past times by the position and velocity at t = 0.

This is not true for atomic systems. As pointed out in the previous section, the Schroedinger equation (a wave equation) describes atomic particles. In Figure 18.4 we try to show that a wave pulse has properties of a particle in that it localizes energy in a small region of space. On the other hand, one must pay a price for this localization. We have already pointed out that a single sine wave of the form

$$E = E_o \sin\left(\frac{2\pi}{\lambda}x\right); \quad \lambda = \frac{h}{p}$$

$$= E_o \sin\left(\frac{2\pi}{h}px\right) \tag{18.16}$$

has its effective mass spread out over space and does not resemble a particle. To make a localized pulse we must add many such simple sine waves over a wide range of momenta. The more localized we want the pulse, the wider must be the range of momenta. So we see that for a localized wave pulse, the position is definite, but that it has many momenta, so the momentum is not well defined. In 1927 Werner Heisenberg (Nobel prize, 1932) expressed these ideas in a form usually called the uncertainty principle.

> The uncertainty principle states that if a measurement of position is made with accuracy Δx and if a measurement of momentum is made simultaneously with accuracy Δp, then the product of the two errors can never be smaller than a number which is roughly the size of Planck's constant h. In mathematical form, Heisenberg wrote

$$\Delta x \, \Delta p \geq \frac{h}{2\pi} \tag{18.17}$$

The Uncertainty Principle reflects the fact that tiny particles with very small masses are significantly affected by measurements. For example, you can get the position and velocity of a moving airplane using radar photons. A radar photon carries so little energy that it does not effect an airplane! On the other hand, suppose you try to detect a moving electron by bouncing photons off it. The electron is so small that it is shoved about by the photons; when a photon bounces off an electron it gives the electron an uncontrollable kick. Your measurement has disturbed the system and some information has been lost. The returning photons do not predict the position and velocity of the electron as precisely as they do the airplane.

Uncertainty principle problems rarely appear on exams, and if they do, they involve straight substitution into Equation (18.17) with no tricks.

EPILOGUE

We end this book by showing you some excerpts from our sequel:

Real Physics Students

Don't Eat Quiche

Real Physics students don't eat quiche!

They eat McDonald's hamburgers -- the regular hamburgers, not the quarter pounders. Real Physics students don't like the high quality beef that goes into the quarter pounder. They like whatever it is that goes into the regular hamburger.

Real Physics students never follow directions on exams. They neither show their work, nor use units. The work and units are obvious:

Page 1

IVY UNIVERSITY

Physics 1   Final Exam

Name (print)

Name (sign)

DIRECTIONS

1. You will not get any credit if you do not show your work needed to arrive at an answer, even if the answer is correct.

Page 2

1. Once every 76.4 years the planets Mercury, Mars, and Neptune line up with the constellation Aquarius. If the lineup occurs during a total eclipse of the sun on earth, what is the velocity of Jupiter's smallest moon at this time?

Answer: 2

INDEX